GLOBAL FLU AND YOU

GLOBAL FLU AND YOU
A HISTORY OF INFLUENZA

George Dehner

REAKTION BOOKS

To Jodi and the boys (Brendan, Patrick, and Sean)

Published by Reaktion Books Ltd
33 Great Sutton Street
London EC1V 0DX, UK

www.reaktionbooks.co.uk

First published 2012

Printed and bound in Great Britain
by TJ International, Padstow, Cornwall

British Library Cataloguing in Publication Data
Dehner, George.
Global flu and you: a history of influenza.
1. Influenza—History.
2. Influenza—Epidemiology.
I. Title
614.5'18-dc23

ISBN 978 1 78023 028 3

Contents

Introduction

"The world is now at the start of the 2009 influenza pandemic."[1]

With these words, World Health Organization (WHO) director-general Margaret Chan officially declared on 11 June 2009 that the new circulating strain of influenza was spreading globally and uncontrollably. For the first time in 41 years, humanity faced the threat of a fast-moving, highly infectious new strain of influenza to which virtually nobody had immunity. "Pandemic," from the Greek *pan/demos*, means "all the people." Anybody and everybody could contract this new flu.

To influenza and public health experts the announcement came as no surprise; in fact, such an event was expected. Some researchers had spent their entire careers studying the virus and preparing to respond to just such a pandemic announcement. This pandemic planning and preparation had accelerated in the last few years as events in the field of influenza research seemed to herald exactly this type of pandemic declaration. To protect the public against the speedy virus required quick action and rapid decision-making. Public health professionals had been busily game-planning responses to a variety of scenarios.

But there was also something wrong about the announcement; or, at least, it was not the announcement that specialists had anticipated. Chan's proclamation did not refer to the projected H5N1 avian influenza strain commonly known as "Bird flu" for which the public health community had been feverishly preparing over the preceding few years. Instead, Chan was declaring the official designation of the Swine flu pandemic of 2009 (subsequently officially called the 2009 Pandemic Influenza A (H1N1)).[2] For the influenza researchers, this pandemic was the wrong strain—an H1N1 instead of the anticipated H5N1—and from the wrong place—originating in North America rather than Southeast Asia. Thus, although Chan's or some other director-general's announcement was expected, this specific declaration still came as a surprise to the scientific and medical community.

Chan's announcement had been expected by the public, too. For the preceding few years they had been conditioned to be alert to and prepare for a pandemic of Bird flu. This message, amplified and indeed often distorted by various media outlets, had cautioned that the world was "overdue" for an influenza pandemic and that a likely candidate—Bird flu—was circulating in Asia, Europe, and even parts of Africa. Usually accompanied by the disclaimer "it may not happen," popular discussions of a potential flu pandemic frequently outlined a series of dire predictions. Therefore, the public was prepared for the arrival of a pandemic of influenza, now recast as an "emerging" disease.

But what Chan's announcement meant was not clear. The informed public had been repeatedly warned in the early years of the twenty-first century that Bird flu was highly virulent. Indeed, as of 7 June 2012, there were a total of 606 confirmed cases of Bird flu, and 357 of these infectees had died. This equates to a mortality rate of nearly 60 per cent, an extraordinarily high and frightening case fatality rate for an influenza infection.[3] Fortunately, thus far Bird flu had not adapted successfully for human transmission. The unfortunate victims had all contracted their infection directly from contact with infected birds.

The pandemic announced by Chan was not the feared Bird flu, however, but a type of swine flu. This new virus had appeared unexpectedly and spread very rapidly around the globe. The first cases were identified on 17 April 2009, and just eight weeks later, the WHO announced that Swine flu was a pandemic.[4] Because the new strain spread quickly, a protective vaccine would not be available for months. Although it had been greatly expanded over the previous few years, the stockpile of antivirals such as Tamiflu or Relenza was only of sufficient quantity to protect a tiny sliver of the planet's population.[5]

Was Swine flu like Bird flu? Worse? Milder? What was the public to do? Unfortunately, nobody could say for sure.

Over the ensuing few months the "meaning" of the Swine flu pandemic of 2009 became clearer. The novel virus spread rapidly in both the southern and northern hemispheres. Vaccine manufacturers labored to produce doses to protect the public. Partly due to production and distribution snafus, but mostly because of the sheer speed of viral spread, the number of people infected with the affliction rose steeply, crested, and began to subside before sufficient stocks of vaccine became available to the general public. Vast stores of vaccine, purchased by major governmental organizations, sat in warehouses and stockrooms because the pandemic had already passed by or the public was uninterested in receiving the protective shots. A significant quantity reached their expiration dates and ended up being destroyed.[6]

Although it remains too early fully to count the costs of the Swine flu pandemic of 2009, it is abundantly clear that overall they were relatively low. Some have projected that the mortality toll was even lower than the average death rate for seasonal influenza.[7] The failure of the pandemic to live up to some of its more dire predictions prompted some skeptics to charge public health authorities with being alarmist or inept. Others of a more sinister bent suggested that the whole pandemic hype had been a plot to boost the profits of pharmaceutical houses and vaccine manufacturers.[8] Even those who were not conspiracy-minded were left with questions in the wake of the pandemic. Where does flu come from? Is it really a big threat to human populations? Should I be concerned about an influenza pandemic, and, if so, why?

This book seeks to address these questions through providing some historical perspective on the pandemic of 2009 and the current concerns with Bird flu. All the events of 2009–10—the rapid spread of the infection, the steep rise and sudden drop in cases, the tardy vaccine production efforts, the stocks of unsold vaccine, and even the charges of conspiracy in characterizing the pandemic—have occurred before. Although the strain that spread epidemically in 2009–10 might be new, influenza pandemics in the human population are not. In fact, the influenza virus has likely been infecting human communities for thousands of years.

As we shall see, influenza is not actually a human disease, or at least was not originally. Instead, it is a bird disease that periodically is transmitted to humans (the technical term is a zoonosis: an animal disease that infects humans). The normal hosts that the virus has evolved to infiltrate, known as its animal reservoir, are the numerous species of aquatic waterfowl such as ducks and geese. At various times in history, the animal virus acquired the necessary evolutionary changes that enabled it to be transmitted to humans and cause infection. The process is fairly complex and will be detailed in the following pages, but it is when this animal disease gains the ability to be exchanged from human to human that epidemics and pandemics emerge. It is the adapted ability to efficiently leap from infected individual to non-infected individual, and the resulting sickness and death the infections occasion, that makes influenza such a threat to public health. When a new influenza virus begins readily to spread between humans, public health professionals, scientists, and medical researchers become concerned because they know something that many of the public do not: influenza is a very costly disease, both in economic and in human terms. It is these costs that prompt officials to go to such great lengths to protect against the disease.

While professionals understand the threats posed by the influenza virus, the general public remains largely unaware of the damage influenza infections can cause. It is for this reason that many people question the exertions of those in public health: the general population has a low level of concern with the influenza virus. Part of the reason for this complacency is that influenza is often misidentified; not by medical experts, but by private individuals. If influenza were a corporate brand name, the first thing its executive board should do is fire its copyright attorneys. The word "flu" has been appropriated to describe a variety of minor ailments ranging from "stomach flu" (a term given for minor intestinal discomfort) to any minor cold or sniffle, and even to the complaints following a night of one or three too many ("beer flu"). A true case of influenza is none of these things. An influenza infection is a contagious respiratory illness ranging in severity from mild to serious, sometimes culminating in death. The course of the illness is generally marked by a combination of the following symptoms: fever, cough, sore throat, runny nose, body or muscle aches, headaches, fatigue and occasionally diarrhea (which is generally more common in children). While it is true that the vast majority of people who catch the illness will recover in a few days to a few weeks, it is also true that some cases may prompt secondary complications that sometimes result in death.[9] The flu has the potential to be much more serious than is generally presumed.

Contributing to the confusion is the swirl of names the influenza virus has gone under in its long history. Sometimes these were drawn from the sudden, dramatic appearance of epidemics or pandemics. In the sixteenth century, Europeans called the affliction *cocqueluch* and *coccolucio*, which loosely translates as "the new fad." In England, this attribute is manifested in the names "the newe acquaintance" and the "new burning ague." Other names were derived from the symptoms of influenza infection, which frequently strikes suddenly. This spawned names such as *l'borion* (a sudden punch between the shoulders) and *la grippe*. In London this characteristic was behind names such as "knock-me-down-fever" or "the knock." In Spanish, the rapid impact is clear in the name *trancaze* (getting hit with a bar) as well as *schnupfenfieber* in German and *mal di castione* in Italian. The relatively low mortality rate also led to more whimsical names, such as "the jolly rant" or "the gentle correction."

Since the late nineteenth century, pandemics have generally been named after where the first infections were identified, giving us the names Russian flu (1889), Spanish flu (1918), Asian flu (1957), Hong Kong flu (1968), and Russian flu (1977). Popular names have also begun to draw upon the source of infectious strains, giving rise to the designations Swine flu (1976), Bird flu (1997–), and Swine flu (2009). The technological ability of today allows us to be certain of influenza outbreaks by laboratory identification, but the sole reliance upon symptomology in the past led to confusion over properly identifying influenza epidemic infections, and hence to a variety of new names, as physicians could not be certain that the affliction they were dealing with was the same as an illness that had previously appeared.[10]

The second action the corporate executives should take to protect their brand name of "influenza" is to fire their advertising firm. Generally, people do not consider influenza to be a dangerous illness. Nearly everybody has had a bout of the flu and emerged none the worse for wear after the illness has passed. In fact, more than 99 per cent of the public will recover from an influenza infection with little apparent ill effect.[11] And, in general, influenza is very selective in who it kills. The burden of mortality falls heaviest on those with weakened immune systems—the very old, the very young and those with some sort of underlying condition that compromises their ability to fend off infections. In fact, influenza was colloquially known as a friend to the physician because many people would get sick and rush to see the doctor (inflating his income), and almost all got better, no matter what the physician did (an important concern prior to the twentieth century, when treatments might include the use of leeches). Influenza's propensity for killing the aged is also manifested in the sobriquet "the old man's friend," as the illness was seen as gently carrying those near death's door across the threshold.[12]

This perception of an unpleasant but generally benign influenza infection misses its full penalties to society. Influenza is an insidious killer whose impact we are just beginning to understand. It may kill a comparative handful of people directly, but infection with the virus elevates mortality in a variety of other categories such as pneumonia, heart disease, and stroke. The Centers for Disease Control and Prevention (CDC) in the United States estimates that each year, seasonal influenza infection causes an extra 36,000 deaths at a yearly cost of $87.1 billion (including $10.4 billion in direct medical costs). Globally, the seasonal flu is estimated to kill between 500,000 and 1 million people each year. That is 1 million mothers, fathers, brothers, children and friends whose lives are cut short because of the influenza virus. And that number is steadily growing as the average age of the global population increases.[13]

It is clear how an influenza infection contributes to elevated case rates of pneumonia. In the process of fending off the viral assault, the human immune system causes quite a bit of collateral damage to the cells that line the respiratory system. The damaged cellular surface allows opportunistic secondary invaders like bacteria to enter into the body, and with the immune system occupied with eradicating the virus, the bacterial interlopers can quickly gain a toehold in infecting the body. Such bacterial colonies cause pneumonia by replicating in the lungs. But the mechanisms for increased rates of stroke or heart attacks are not so clear. Statistically, it is easy to see that these rates climb in the wake of influenza's visit to a community, and this increased mortality rate that trails influenza outbreaks has been known for over 150 years, but the precise interrelation between the two events remains something of a mystery.

The costs and deaths discussed thus far are related to seasonal influenza, and, as we have seen, seasonal influenza is no minor event. But this yearly impact of the virus—sometimes greater, sometimes lesser—is not really disruptive to society. A bad flu year may stress a region's healthcare system, but it rarely overwhelms it. The possibility of overwhelming the medical infrastructure is a distinction reserved for pandemic influenza. With its sudden, steep rise of the infected, the rapid accumulation of the sick in hospitals and medical offices, and, in some cases, the increased number of dead, it is pandemic influenza that becomes so difficult for communities to handle. Even a mild pandemic can overstress medical systems because of the large number of people who contract the virus at once. These ill people flood into their doctors' offices and emergency rooms seeking treatment, creating a surge that is difficult for health systems to cope with. This surge occurs even if the pandemic has low mortality rates, and, as we shall see, not all pandemics have been so mild. For these reasons, protecting against and preparing for pandemic influenza has been the central focus of public health in terms of influenza.

Pandemics are random events and occur when the stars of novelty, human infectiousness, and efficient transmission line up for an influenza strain.[14] When these three elements align, a pandemic is born. Notably, virulence is not part of the pandemic equation. Virulence amplifies the impact of a pandemic, but it is not necessary for its onset or spread. And historically, there have been mild pandemics (as apparently the Swine flu of 2009 was) as well as very virulent ones (as was the case for the Spanish flu of 1918). For that matter, virulence may decrease or increase over the course of a pandemic and, despite the power of modern genetic tools, scientists remain unable to predict whether a circulating strain will be of high or low pathogenicity. In addition, there is no sure predictor of how any individual will react to infection with the virus. Two people infected with the same strain may have dramatically different courses of infection. One person may have an infection so mild that they are asymptomatic, meaning that over the course of their illness they do not manifest any symptoms. However, even an asymptomatic individual can be infectious. Someone with a very mild course of infection can transmit that virus to someone who could have a far direr, even deadly, bout of the flu. These individual responses are maddeningly unpredictable. There are also cases in which there is a lag between the host's ability to transmit the virus to others and the appearance of his or her own symptoms. It is for this reason that quarantine efforts, such as scanning incoming travelers for fevers at airports or border crossings, are doomed to fail in halting the spread of a pandemic. Though the percentage of asymptomatic infected people is impossible accurately to estimate, it is generally presumed to be a significant minority of cases.

Unpredictability remains a constant in studying the influenza virus. As we shall see over the course of this book, influenza has had a long shared history with human populations. Changes in lifestyle, living patterns, transportation, and population size have marked the transference of the virus from its animal hosts to human communities. The change from nomadic to settled life brought larger human populations into contact with this disease of birds. Domesticating birds and other animals created mechanisms that increased the ease of transfer of the virus from animal to human host. New communities grew and expanded, interacting with other growing settled communities to create a very large, interconnected population. This altered model of human environments facilitated the creation of new human diseases from diseases that had formerly been found among animals; influenza was among them. Pandemic influenza was born.

Influenza epidemics have surely emerged and swept through populations with greater or lesser impact for centuries, as have a myriad of other terrible infections. But understanding the disease and the reasons for its appearance, impact, and disappearance remained beyond the comprehension of our ancestors for an equally long time. Simply put, if you think that sickness is caused by angry gods, the influence of the stars, bad vapors or some combination of these and other explanations, it is unlikely that you will be able to develop effective medical techniques against fast-moving diseases. In addition, in the case of influenza, it was often difficult to fully appreciate how many the virus killed each year against a backdrop of a truly appalling burden of infections that kept mortality consistently high. Identifying influenza pandemics in the past from cryptic references in ancient chronicles and manuscripts is something of a parlor game for medical researchers, but we can say with a high degree of confidence, based on the new living patterns and animal–human interactions, that influenza was present.

The living patterns set in motion by the Neolithic Revolution have continued since our early farmer ancestors, and in the past few centuries have greatly accelerated, so much so that they constitute a revolutionary living arrangement change almost as profound as the abandonment of nomadic life. Human populations are increasingly urban citizens, and this urbanization process has been greatly magnified by soaring population growth. In addition, burgeoning global cities are connected by a speed of transportation and a volume of trade and travel that are unprecedented in human history. The global population is now more primed than ever for the rapid transfer of highly infectious diseases. Fortunately, in the twentieth and now twenty-first centuries, scientists and medical researchers have made great strides in understanding, treating, and preventing the spread of diseases, including influenza. In some ways, we are better prepared to fend off an influenza pandemic than at any previous time in human history. Greater understanding of the causative element and new tools and techniques offer the chance not only to protect against epidemic spread, but potentially to forestall the emergence of a pandemic. But in other ways, our new, greatly interconnected world makes us more vulnerable than ever. For the immediate future, we remain under threat from this unpredictable entity.

Understanding the long history of interaction between humans and the influenza virus is important because it will help us both to prepare for the next pandemic and to see its danger in perspective. Influenza's comparatively low mortality rate pales in comparison to those of other epidemic diseases. But, lest we grow too complacent, influenza history also records the Spanish flu pandemic, which killed somewhere around 50 million people in only a few months—likely the most deadly pandemic in the shortest time in human history. That is emblematic of the confounding nature of the virus; it can produce pandemics that kill a comparative handful, like the Swine flu of 2009, and ones that can kill millions. We ignore influenza at our own peril.

ONE

Know Your Enemy

Before we begin to discuss the impact of influenza in the human population, we need to clarify what it is, how the human immune system responds to it, and where the organism comes from. As we shall see, the use of the term "influenza" to define a specific sickness has evolved over time, as has the recognition of the disease's impact. The material below is sketched out from the latest research about the virus and the disease it precipitates. The following discussion is mildly technical, but it is crucial that we clearly understand the virus and the reactions it prompts. Therefore we will begin at the beginning by describing the virus and detailing how the human body reacts to fend off infection.

The Virus

Influenza is caused by a virus from the family Orthomyxoviridae, from the Greek *orthos*, which means "straight" or "correct," and *myxa*, meaning "mucus."[1] There are three types of influenza virus – A, B, and C – named for the order of their discovery. Type C is a rare kind that elicits a very mild infection in the human body, and apparently can only naturally infect human hosts. Type B is also a primarily human disease that can spread epidemically, prompting illness, complications, and even death. In general, influenza B strains occasion milder outbreaks that do not transit as explosively as influenza A infections. Type C is a curiosity, but type B can be dangerous to human populations. However, the threat of type B influenza infections is greatly eclipsed by the dangers an influenza A epidemic poses. Therefore, this book will discuss the influenza A virus unless specifically noted otherwise.

Viruses are unique organisms that have prompted some scientists to question whether they should be classified as "living" entities. The dispute stems from the fact that viruses cannot replicate without infecting a living cell. Aside from having a strong drive to reproduce themselves, they do not possess any of the other attributes generally associated with life. Viruses are simply pieces of genetic code encased in a protective shell that must enter another organism in order to replicate and thus propagate themselves for the next generation. Without this hijacking of other organisms, the entity is inert. Viruses are therefore invaders and parasites.

The influenza virus is a fairly simple snippet of genetic programming, but the characteristics of the code and the structure it creates provide it with a fascinating bag of tricks. The virus is comprised of eight genetic segments, which code for ten proteins that create the internal structure of the virus and encase it in a protective membrane. It should be pointed out that despite decades of close study utilizing a variety of powerful tools, scientists are still not sure of all the functions of the eight segments of the virus's code. That said, over the years researchers have been able to determine quite a bit of information about the genetics of the virus and the functions of the proteins it creates. Because the components of the viral shell are so important for both its replication ability and the human immune system's response, the genetic elements that create them have drawn intense scientific focus. Jutting out from the surface of the virus are two glycoprotein appendages, which play a role in entering and exiting the targeted cells necessary for reproduction. The first, termed hemagglutinin (rendered as H), forms roughly 90 per cent of these protrusions and is generally pictured as being spike-like. The remaining 10 per cent of the surface projections are produced by the neuraminidase (rendered as N) section of the genetic code and are often described as mushroom-shaped. Hemagglutinin plays a crucial role in entering a living cell while neuraminidase serves to cut loose the daughter copies that bud out from infected cells, thus setting these descendants free to infect other cells and organisms. These spike- and mushroom-shaped elements of the virus's appearance are clearly visible under electron microscope magnification.

When scientists speak of "influenza," they are actually discussing a generalized group of closely related viruses. In addition to the three types of influenza (A, B, and C) already discussed, influenza A has a number of subtypes, grouped according to the distinctive characteristics of the hemagglutinin and neuraminidase. There are sixteen different types of hemagglutinin identified by researchers (labeled H1 to H16) and nine types of neuraminidase (N1 to N9). An influenza virus is identified by its combination of hemagglutinin and neuraminidase components. So, for example, the Swine flu strain of 2009 was an H1N1: a hemagglutinin type from family 1 combined with neuraminidase type 1. The Asian flu of 1957 was labeled H2N2 and theoretically there could be a virus combining any one of the sixteen hemagglutinin and nine neuraminidase families. Until the early twenty-first century, it was generally considered that only H1, H2, and H3 and N1 and N2 types were infective for humans, since these were the only hemagglutinin and neuraminidase kinds associated with human influenza infections. But the human illnesses and deaths associated with H5N1 Bird flu as well as the discovery of H7N7 and H9N2 infections have invalidated this prevailing hypothesis.

The hemagglutinin, the neuraminidase, and the remaining six segments of the virus's genetic make-up (PB1, PB2, PA, NP, M1 (2), and NS1 (2)) are contained on a single strand of RNA. Fundamentally, all the influenza virus seeks to do is to get this simple strand of genetic blueprint into a living cell in order to make copies. To achieve that goal, the virus first must evade the immune system of its targeted host. As we shall see, the specific attributes of the influenza virus enable it to be successful in this primal drive of genetic survival. But before we can discuss how the virus evades the immune system, we need to understand how the immune system works.

The Human Immune System

The human immune system is a wonder of evolutionary engineering.[2] In response to a relentless barrage of viral, microbial, fungal, and parasitical invaders, the human body has developed a multi-level integrated defense network. At the most basic level, the body seeks to prevent entry by outside entities. My entire body is enveloped in the protective barrier of the human skin, a surprisingly sophisticated structure designed to prevent contamination. This protective covering is of course broached by a number of egress and ingress points necessary for human functioning and all can serve as gateways for potentially dangerous outside matter. No bodily function is more potentially fraught for infection than the act of breathing, whereby air from the outside environment is drawn deep into the core of the organism, where the delicate exchange of oxygen for carbon dioxide occurs a dozen or more times a minute. To intercept potential breath-borne hitchhikers, the human body has a series of defense mechanisms designed to catch alien material. The passageways of the breathing system are lined with hairs and a special substance known as mucus is produced, which captures and binds up foreign objects inadvertently dragged into the body. These alien entities are neutralized by digestion or expelled by coughing and sneezing.

The actions of this physical response capture the foreign matter before it can cause any ill effects in the organism. But even the most effective prevention system cannot be entirely foolproof. Infectious material may elude capture at the entryways to the organism, or the protective barrier of the skin may be pierced by cuts or jabbing, which may insert outsiders into the body. Therefore, this initial line of defense is backed up by a series of patrol cells that circulate throughout the bloodstream. These first responders are named the innate or non-specific immune response, because their reaction is not tailored to the specific invader. These cells possess the ability to detect "self" from "non-self" among the things they encounter. When a type of cell known as a phagocyte detects something it identifies as non-self, it captures and envelopes the intruder, releasing a powerful enzyme that destroys and digests the foreign matter. The relentless focus of these powerful cells, which continually circulate through the body, provides a very efficient defense mechanism against trespassers.

To boost the effectiveness of these non-specific defense cells, the body has evolved practices that enhance the ability of the non-specific immune system to respond in numbers to larger invasive threats. When the body suffers an injury such as a cut or a blow, the organism reacts by generating localized swelling, whereby fluids are forced to the site. This both restricts movement out of the affected area and rushes in more phagocytes. In addition to phagocytes, the innate system contains another first responder, known as a natural killer cell. Natural killer cells can detect non-self from self, and can also identify when a self item, like a cell, is infected by a non-self entity. When a natural killer cell finds an infected cell, it releases a powerful toxin that destroys the co-opted cell. The ability to mark and destroy cells infiltrated by an alien force is particularly useful in fending off viral infections since, as we have seen, a virus needs a living cell in order to make a new generation.

As the innate immune system goes about its business of eradicating foreign material, chemicals are released into the bloodstream. When a threshold level is reached, the hypothalamus is induced to raise the body's temperature. At higher temperature levels, the innate immune system operates more effectively and viral and microbial replication is hindered. This first system is often sufficient to defend the body and evict or destroy the offending substances, but not always. Sometimes the infectious material is too plentiful or too effective at generating more offspring and the primary immune response cannot keep pace with it. Fortunately, the human organism has developed a second line of defense, one that is keyed to specific invaders and provides a lasting capability to prevent future infections.

This second, sophisticated type of protective response is termed the cognate or specific immune reaction and is tied to the distinguishing aspects of the outside appearance of the foreign body. The initial innate immune system response is effective in neutralizing pathogens, and the unconscious body responses of swelling and temperature rise do help slow down replication by the alien material. But the speed of the innate system is outstripped by the incredible growth potential of outside organisms, especially the replication ability of viral invaders. In the specific case of an influenza viral infection, cells infiltrated by this interloper can be churning out copies as quickly as four hours after being infected by the virus. Within 24 hours, a single influenza virus produces over a billion copies.[3] Such high replication rates simply overwhelm the first level of immune response.

Therefore, the specific immune response must be able to match the exponential rise of the viral invader. The solution to this quandary lies in special cells produced by the thymus (T cells) and in the bone marrow (B cells). The thymus and bone marrow are continually at work over the entire duration of a person's lifetime generating these second-level immunity components. The B and T cells possess the ability to bind with the outside structures of foreign entities, thus preventing them from attaching to cells to initiate viral reproduction. B and T cells that are effective in latching on to the outside structures of foreign material are known as antibodies, and they are formed as a part of the body's interaction with infectious material.

The process unfolds in this fashion. As the non-specific phagocytes and natural killer cells dismantle non-self entities, fragments and pieces of the destroyed trespassers are generated. Pieces of this debris are picked up a specific cell type known as an antigen-presenting cell, which transports its cargo to the lymph nodes. The lymph nodes are packed with a variety of T and B cells. If the body has encountered this infectious matter before, it retains a record of that infection in the form of a memory B cell, which is precisely molded to the outside appearance (or portion of the outside appearance) of the invading organism. The reappearance of the infection prompts the memory B cell to massively produce copies of the antibody. These flood into the bloodstream and rapidly attach to the targeted outside features of the foreign material, preventing it from binding to a cell and slowing it down for capture and destruction by phagocytes. The antibody production bolsters the front-line defense and the two systems (innate and cognate) combine to destroy the interlopers.

This ability of the body to recall previous infections through the creation of memory B cells prevents re-infection with alien material and is what we call immunity. The production of antibodies is particularly effective against viral invaders. The organism's ability to prevent the binding of the virus to a host cell foreshortens or prevents infection. The rapid and effective response detailed above illustrates how the body responds to the reintroduction of an invasion of foreign material; but how does the immune system respond to a novel infection? The answer again can be found in the collection of B and T cells located in the lymph nodes.

In addition to retaining memory B cells keyed to every infection the body has endured before, the lymph nodes boast a variety of differently shaped, non-specific (or generic, if you will) antibodies. When an antigen-presenting cell brings recovered pieces of foreign organism to the lymph nodes, T and B cells that are close in outline to the exterior shape of the alien entity are stimulated to make copies. The body is then flooded with a variety of antibodies that are more or less successful in neutralizing the infectious matter. The closer the antibody match to the target, the more the body is stimulated to make copies of that analogue. Over time a perfect antigenic mold is selected. The lag between the introduction of an invader and the massive production of antibodies to interdict the invader is called the infection. The success in selecting and producing the perfect template eradicates the infection. In the aftermath of the contagion, the body is swarming with antibodies prepared to bind up the reappearance of the alien and prevent re-infection. Over time this supply of antibodies dwindles to just one memory B cell, which is stored in the lymph nodes to await the return assault of the infectious foe.[4]

The effectiveness of the specific immune system accounts for the fact that re-infections by the same virus are rare, and also explains how an affliction like smallpox can be successfully eliminated.[5] But it sets up a conundrum in understanding influenza infections. We know that influenza continues to circulate in greater and lesser volume each year, and that a bout of the virus does not mean you will not get influenza again. How, then, is the influenza virus able to evade the powerful human immune system?

Influenza's Evasive Tricks

The ability of influenza to circulate and re-infect people lies in some unique aspects of the virus's genetic code.[6] The first explanation for the virus's skill in evading the human immune system resides in the fact that influenza is an RNA virus. When the influenza virus invades a living cell, it compels the cell to recreate the letters that comprise the genome of the virus. Unlike a photocopy, however, generating a replica of the viral code is a process with the potential for errors. Mistakes arise all the time, but certain types of reproduction make copying blunders more likely. RNA viruses have a much greater capacity for genetic mutations than DNA genetic copying. These mutations are the result of transcription errors. Unlike DNA replication, RNA copying lacks the code-checking mechanism that ensures that the reproduced code is a match for the genome of the parent. To use a simple analogy, RNA reproduction is like typing without proofreading. In recreating the viral code, a number of copying errors (or "typos") are introduced into the facsimile being created. Left unchecked, these alter the genetic offspring. DNA replication includes a step that matches the copy with the original source code in which errors are detected and corrected. RNA genome creation lacks this mechanism. In general, this means that mutations arise far more frequently in RNA reproduction when compared to DNA copying. In fact, there are about 1,000 mutations in RNA viral creation for every one DNA mutant. Of all the RNA viruses, the influenza virus is one of the most prolific mutant producers (or, if you prefer, one of the most error-prone typists). The transcription errors can impact any or all of the eight genetic segments of the influenza virus, and perhaps 10 per cent or more of all genetic offspring of influenza viruses contain some changed viral text.[7]

Among the many deviations produced by influenza virus copying are ones that impact the arrangement of the hemagglutinin and neuraminidase projections that stud the outside of the viral shell. These errors result in the subtle rearranging of the pattern of spikes and mushrooms. This changed appearance of the outside of the virus thwarts the specific immune system response that, as we have seen, develops antibodies to latch on to the outside of viral invaders. Mutant daughter cells therefore evade capture by the specific immune system response and so are able to continue to replicate and invade other cells in the body, or are available to be expelled into the wider environment and thus infect a new host. This rearranging process due to altered genetic mutations is known as "viral drift," as selection pressures favor slightly altered viruses. Over time the collective changes continue subtly to reshape the virus, ultimately resulting in a virus that appears markedly different from the original source material and to which a previously exposed individual no longer possesses immunity.

As effective as viral drift is in evading immune system response and continuing its human-to-human exchange, the percentage of people vulnerable to the slightly changed virus is limited. Because the viral shell is only a little different from the strain encountered before, some people's immune systems are quickly able to adapt to the new shape. Over time, the continual circulation of the viral type results in a variety of antibody structures that offer cross-immunity to the newer viral pattern. The resistance to this influenza type tends to increase the longer the virus circulates, therefore making it more difficult for the infectious material to encounter susceptible people to infiltrate. The cumulative effect is a dampening of influenza infections. Yet such a viral drift process cannot account for the rapid, nearly universal, rise of influenza outbreaks that marks a pandemic. Where, then, do pandemic strains come from?

The answer is again a feature of the virus's unusual genetic structure. The influenza virus's RNA code is comprised of eight segments, or packets, of information that produce the various components of the virus. When a virus invades a host cell it coerces the cell to make copies of each of the portions of the code, which are then re-knit to produce an intact viral code. If an organism is infected with one type of virus, the invader turns the infiltrated cells into little factories producing more or less exact copies of the parent virus. However, if a person is infected with two different influenza types (say an H1N1 and an H2N2), the infected cell serves as not only a factory but a kind of conjugal bed. The influenza virus has a proclivity to swap its genetic packages when stitching together the eight viral segments of its genetic code. If a cell is co-infected, this swapping process can result in a virus that contains a mixture of the packets of two different viral strains. So a virus produced in this co-infected cell could take some genetic segments from virus "X" and combine it with some from virus "Y." The result is a brand-new virus "Z" that marries elements of parent viruses "X" and "Y." This process is how pandemic strains are formed, and is known as a "viral shift" because the virus has shifted its shape dramatically. This combination process can certainly occur in a human host, but for reasons that will become clear shortly, it is more likely that an animal host will serve as the bordello for these new, shifted viral types.

Thus far we have been discussing influenza as a human disease, and for the bulk of this narrative that will be the case, but it is important to recognize that influenza is not a human disease; or, perhaps more accurately, not originally a human disease. A variety of animals can contract influenza, including horses, dolphins, whales, pigs, and others, in addition to humans. But the animal reservoir and likely original hosts of influenza viruses are birds; specifically aquatic waterfowl such as ducks and geese.[8] All sixteen hemagglutinin and nine neuraminidase types have been recovered from these bird populations.

Influenza strains in waterfowl, although fundamentally the same virus that infects humans, have some important differences from those in human populations. First, in the bird population influenza is an alimentary affliction, meaning it infects the digestive tract and is shed by excretion rather than by respiration. Second, an influenza infection in these bird populations apparently causes them little harm. The birds' symptoms generally manifest as some ruffled feathers and decreased feeding, and do not hinder their ability or willingness to fly.[9] The combination of these two features means that infected birds travel widely, depositing the virus in the lakes and streams they linger over. In addition, the virus remains viable in the environment in the cold-water watersheds that migratory birds favor. Infectious viral strains have been recovered from these waters weeks after migratory populations have moved on, and presumably infect other transients. The low level of impact on an individual bird's health and the relative ease of dispersal of the virus into the wider environment means that birds are frequently exposed to influenza infections. Indeed, studies have shown that at certain times of the year up to 25 per cent of juvenile fledgling populations are harboring an influenza virus.[10] Influenza infections are ubiquitous in this bird population.

The benign nature of the infection, the ease with which the virus infects the bird host, and the fact that all types of hemagglutinin and neuraminidase have been recovered from waterfowl populations suggest that influenza has been a bird affliction for a very long time. Bolstering this supposition is the fact that in these bird populations, the virus changes very little over time, in sharp contrast with the high mutability rates in human populations. In fact, the viral types can almost be termed as being in "genetic stasis." This absence of viral drift is not because the RNA replication does not create mutant offspring—the mutation rate of the RNA virus remains the same— but because the virus is so supremely adapted to transmission in the duck and goose gatherings that new genetic incarnations of the viruses do not offer a competitive advantage against the prevailing strain and so are not selected for continual transmission. Because influenza transits from host to host so readily among birds, the stray mutations produced do not have any reason to supplant the viral types already circulating.

Since influenza is also a human disease, the virus must have become adapted to infecting humans as well as birds. Unraveling the process of turning this bird illness into a human affliction has two main components that need to be understood. One is mechanical: How does a virus that infects the intestines of birds become one that infects the respiratory tract of humans? This question will be addressed below. The second question about human infection is more broad-based. How did humans come to inhabit a shared living space with these animals and so come into close contact with this infectious agent? This question will be taken up in chapter Two.

Cellular Infection

In order to understand how influenza becomes a human disease, we need to clarify how the virus infects birds at the cellular level: specifically, how the virus induces a living cell to transport this alien substance into itself. Waterfowl are continually exposed to influenza viruses because they are excreted out by infected birds. When a duck ingests some lake water contaminated with influenza virus, the viral particles travel into the intestines of the bird. In this fashion the virus is introduced into the environment it needs in order to replicate. As we know, to make new copies, the virus must hijack a living cell to do its bidding. This is not as easy as it sounds. In order to operate, cells separate themselves from their surroundings by enclosing themselves in a viral membrane. This border keeps the cell's organelles in and other things out. But cells also have to communicate with their neighbors, and so the cell membrane surface contains various portals that accept messenger chemicals from a variety of sources in the body. On the cells that line the intestines of birds (known as epithelial cells) is a chemical receptor site known as sialic acid alpha 2,3. This site is the entryway that the influenza virus exploits in order to gain entrance to the cell. The tool the influenza virus uses to bind with the cell is one of the two surface components that extend out from the viral shell: hemagglutinin. The hemagglutinin spike on the surface of the virus attaches to this portal like a key in a lock and induces the cell to take the virus into its interior. Here the virus begins to turn the cell to the purpose of replicating the viral code.

The process in the human organism is very similar to the avian process. In the human body, the respiratory tract is lined with epithelial cells, and these cells also have receptor sites for sialic acid messenger communication. But in humans these portals (termed sialic acid receptor alpha 2,6) are shaped differently so the hemagglutinin spike of an avian strain does not latch onto the cell. In effect the avian hemagglutinin "key" does not fit the human alpha 2,6 "lock." Without this connection the virus cannot enter the cell and make copies, and thus cannot infect the human host.[11] Therefore, in order to become a human disease, the virus must adapt its spike to link with the human epithelial cell.

It was generally believed that some unidentified intermediate source was needed to bridge the transformation from an avian disease to a human disease because the sialic acid receptor sites are so different. However, recent research connected with the recovery of the virus that caused the infamous Spanish flu suggests that it is possible that the influenza virus could generate a spontaneous mutation or mutations that allow for the direct infection of the human host with an influenza virus that has predominantly avian characteristics. Still, if this depiction of Spanish flu is correct, the relative lack of genetic pressure on avian strains of influenza suggests that this is a highly rare occurrence, and the virus's unusual avian features may account for the ferocious and unusual mortality pattern of Spanish flu. The Spanish flu case aside, influenza virologists generally agree that there likely is some sort of host that serves as a waystation or a midwife in the birth of new human influenza strains;

some organism, or organisms, that serves to shape the virus on its transition from enteric bird infection to respiratory human infection. The most likely candidate for this role is the pig.

Pigs have been proposed as a "mixing vessel" for new influenza strains.[12] Like humans, pigs have epithelial cells that line their respiratory tract and these cells have chemical receptor sites for sialic acid messages. But unlike birds and humans, which generally have alpha 2,3 and alpha 2,6 types respectively, pig epithelial cells contain both kinds of receptors. What this means in practical terms is that pigs are susceptible to catching both avian and human strains of influenza. Thus they can serve as the transit point for avian strains of influenza to infect humans in two ways. First, having alpha 2,6 receptors on the cell surface does offer a competitive advantage to mutant avian strains that develop the ability to bind to both alpha 2,3 and alpha 2,6 sites. The increased number of binding sites means that a mutant avian strain that can attach to both an alpha 2,3 and alpha 2,6 receptor can outproduce an avian strain restricted solely to alpha 2,3 sites, thus resulting in selection pressure on the virus. Second, the dual receptor sites readily allow for simultaneous infection with an avian and a human influenza strain. Because of the virus's affinity for swapping genetic packages, such a twin infection will produce a number of offspring that are hybrids of the avian and human strains. These melding together of two influenza strains from two different species actually results in a spectrum of dramatically different viruses. The vast majority of these mutant hybrids are not viable in one fashion or another. But viruses that are both radically different and able to easily transmit between human hosts are highly likely to become pandemic strains. This swapping of genetic attributes between bird and human strains inside a pig is likely the cause of the pandemics of 1957 and 1968. In the H1N1 Swine flu pandemic of 2009, the virus was the result of an unusual triple reassortment combining avian, human and two different kinds of pig strains of influenza.[13]

The new, hybridized viruses had never been encountered by humans before, so the population was almost uniformly susceptible. The novel viral types spread widely, as each new host lacked immunity to the strain, and each person could serve as the locus of an infection to others. As the newborn viral types spread, repeated passages through a chain of people honed the ability for human infection, making the virus increasingly better adapted to its new host. The novelty of the affliction and its efficiency at transmission makes this illness one that afflicts *pan demos*, or all the people. In this manner the bird infection has now become a human infection.

The influenza virus, like many viruses, is basically a simple piece of genetic code. If the virus was stable, a handful of infections would protect a person throughout their entire life. But the virus is not stable; indeed, it has been called a "slippery" disease.[14] This high mutation capacity and its unusual segmented genetic code combine to enable the virus to continually evade human immune systems and periodically spark pandemics. But this virus is only a comparatively recent invader of the human population, compared to the thousands or perhaps millions of years over which it has been infecting avian populations. The key to the creation of the human disease of influenza required the close interaction of human, bird, and probably pig populations. The unintended creation of human influenza infections was the result of these three species coming together as a result of a revolution: a Neolithic Revolution, that is.

TWO

The Murky Past of Influenza

The very bushy human ancestor tree that evolved in Africa created a number of proto-human species of increasing cleverness and sophistication.[1] Periodically some members of these smarter hominids (Homo erectus about 1 million years ago, H. neanderthalensis about 250,000 years ago, and perhaps others) migrated out of Africa and extended their range across the Eurasian landmass. Over time, the evolutionary cauldron of Africa produced a new hominid species, H. sapiens sapiens, which emerged around 150,000 years ago. This big-brained species eventually moved out from the African continent, supplanting the earlier travelers and their descendants. Eventually, these new hominids colonized the entire planet, having developed techniques to survive harsh climates and to travel across open stretches of water. By about 12,000 to 15,000 years ago at least, these modern humans had populated every continent save Antarctica.

Remarkably, these bright hominids—with an increasingly refined toolkit, complex social and cultural traditions, and an inventive and inquisitive intelligence—pursued the same survival strategy of hunting and gathering that had sustained the earliest proto-human species for millions of years. Of course, over this time frame, the menu was greatly expanded and purposeful hunting came to replace opportunistic scavenging, but fundamentally, food acquisition remained the same for over 99 per cent of human and their ancestors' history.[2] Such a lifestyle carries with it inherent constraints. First, the size of a hunter-gathering grouping is necessarily limited by the resources available. Bands that grew too large placed strains on their food sources, threatening the continued existence of the group. The largest of such bands resided in resource-rich places like the Pacific Northwest, where abundant seasonal runs of fish stocks provided a cornucopia of food sources; and the smallest groups were those that inhabited the harsh outback of desert environments like Australia, or the unforgiving climate of the Arctic. Still, although those with abundant natural resources could support a significantly larger population than those without such a bounty, food availability placed a boundary on population size in all these environments. Second, this living pattern necessitated movement, either to track mobile prey or to visit a variety of plant sources that ripen at different times. This shifting settlement style occurred within clearly defined regions; at least, clearly defined to these peoples and their neighbors. Therefore, the area a bonded group relied upon for sustenance had to be fairly large, and was jealously guarded.

Combined, these two features of hunter-gatherer life—limited population size and varying degrees of mobility—meant that humans and their ancestors resided in comparatively small and isolated groupings. Certainly, this isolation was not pristine: these bands bumped into one another as they tracked game or gathered foodstuffs, and as human culture became increasingly sophisticated and complex there were likely larger gatherings of people who came together for feasts and festivals or to trade and barter. But the seasonality of plenty made these larger population assemblies intermittent.

These human extended family units must have been afflicted by a number of parasitic entities (such as worms, flukes, and other fellow travelers); slow-growing illnesses with a long potential for infectivity (as in yaws or leprosy); and infections that can infect a variety of species beyond just humans (malaria, for example).[3] Afflictions that were highly transmissible or whose course of infection was rapid in an individual host would find humans inhospitable. Fast-moving infections would cycle rapidly through a small population, quickly ending the chain of transmission because the afflicted would either succumb to the illness or develop immunity and so prevent re-infection. The relative isolation of these groupings meant that the opportunity to infect another collection of humans was a comparatively rare event. In addition, the mobility of the population meant that they were constantly leaving their garbage and wastes behind when they moved to new places, which also served to limit the infection rate.

Such a living pattern does not mean that these early human populations were not exposed to infectious diseases. It is a near certainty that human contact with the natural world served to introduce novel infections into the population, but the limited size of groups and the infrequent contact with others meant that these new contagions had little opportunity to persist in the human species. The hunter-gatherer lifestyle provided protection against new, introduced diseases in general, and especially the type that would later be termed epidemic diseases. If such an introduced agent proved to be either particularly nasty or highly infectious, it was quickly extinguished before it could continue to transmit, and thus disappeared. New human living patterns would have to emerge to disrupt this ancient human–disease interaction.

The Neolithic Revolution

A new living arrangement did appear: the Neolithic Revolution. This radical new settlement pattern set off dramatic changes that still reverberate to the present day. Simply put, this revolution turned hunter-gatherers into farmers, and farmers lived in dramatically different ways from their nomadic forebears. The key to the transformation lay in the domestication of plants and animals, through which human masters turned these organisms into their servants. This new food production system did not occur overnight; rather, it was an evolving process that happened over thousands of years and in a variety of places. The end result was an altered living arrangement that changed small, mobile human groupings into settled, expanding, and concentrated populations. Being settled meant remaining in one place among accumulated waste and garbage and among the parasitic creatures that thrived in that disorder; the expanding and denser populations meant that humans were in closer contact with increasing numbers of people. This new living process produced a storable surplus of food, which was the key ingredient for the development of advanced technological, cultural, and societal structures we collectively call "civilization." Such a revised sustenance method also created a new disease environment, one in which the newly domesticated animals played a prominent part.

Domestication brought humans into closer contact with animals than they had had in any previous society. Certainly, humans had been in close contact with animals previously; after all, to hunt one must get close enough for the kill, followed by the butchery and consumption of the prey. But domestication brought humans into proximate sustained interaction with their animals. Humans literally lived with their animals, and not just single animals, but herds and flocks of them. This sustained intertwining with large numbers of animals meant exposure to a number of animal pathogens. Humans were not the only organisms infective agents had evolved to prey upon. Close interaction provided an avenue for any animal disease to make the species leap and infect a human host. To some degree, such a species transference had surely occurred in hunter-gatherer and animal connections. The crucial difference in this post-Neolithic world was that a species jump into this new human host now offered additional opportunities for subsequent transfer.

Humans were not only living in one place as a result of this new food production system: they were also living in greater numbers and in closer contact. Whereas a new infection introduced into a mobile isolated band would quickly burn through this limited population, resulting in either death or immunity, new afflictions introduced into this settled world would find a number of hosts to infect. Continual human transfer produced new evolutionary stresses on the pathogen, reshaping the organism for effective human transmission. Over time, the multiple passages in human hosts transformed these former animal diseases such as distemper or the pox family into the human diseases of measles and smallpox. Measles and smallpox are classic examples of settled diseases because they simply could not have existed in previous times. The diseases are so infectious, and survival produces such long-lasting immunity, that the infectious agents need a certain density of population to pro-duce new susceptible victims in order to maintain transmission. These so called "crowd diseases" would quickly cycle through the small populations of hunter-gatherer society and the necessary chain of infection would end. Epidemiologists often use the analogy of a fire when referring to infectious diseases, and it is particularly apt. An infectious agent is like a spark that needs continual inputs of fuel to keep burning. When the fuel is burned up, the fire goes out. Non-immune humans provide the wood to keep the fire of a pandemic burning.

Against this generalized backdrop of a transformed world that facilitated the emergence of cross-species infections and the persistence of these new afflictions, we can now focus on the two species that are crucial to understanding the human disease of influenza: the duck and the pig. The duck (or perhaps the goose or swan) as having been the animal host of the influenza virus is a given among influenza researchers. As we have seen, ducks host all the known family types of hemagglutinin and neuraminidase and the virus circulates readily through their flocks with few apparent ill effects. And ducks and waterfowl must have been a tasty addition to the human larder for a very long time. But discovering when these animals became captive members of the barnyard is difficult. It seems highly likely that geese have been domesticated for at least 4,000 years, but it is not clear if ducks were also similarly transformed from their wild ancestors at the same time.[4] Part of the delay in domesticating ducks (if there was one) may stem from the fact that ducks are so easily captured. Whole flocks can be netted in their watersheds and breeding areas and the eggs can be readily acquired from nesting sites for immediate consumption or for the rearing of chicks.[5] Rather than being a new, domestic species created by a one-time transformative event, ducks, like other domesticated species, were shaped by a process of human selection over centuries. Gradually these animals have been changed from wild to the domestic species they are today. Since domestication is a continuum, it remains difficult for scientists to conclusively state whether bones are from a wild species or a domestic one; this challenge for archeologists and anthropologists is especially true when examining the bones of ducks.

In contrast to the history of duck domestication, tracing the domestication of pigs is comparatively easy. Simply put, domestic pigs look different from wild pigs. The process of domesticating pigs results in morphological changes most clearly evident in the teeth and bone structure of these animals, which are visibly different from the bones and teeth of wild boars. In addition, farmed pigs tend to be consumed at a younger (more tender) age and this consumption is weighted towards male pigs, as sows are kept for breeding purposes. These distinctions of bone and tooth shape and a preponderance of male remains are apparent in the bones collected from the middens of early human settlements. But the domestication of pigs should also be seen as lying along a continuum, whereby smaller pigs kept as herds in the barnyard are the end result of a larger process of human/swine interaction.[6] Again, rather than being a one-time event in one location, the evidence suggests that pigs were domesticated in a number of places, with regions of East Asia among the earliest to succeed in this endeavor. There is evidence to suggest pigs were fully domesticated between 7000 and 9000 BCE in southern China, by 6500 BCE in the Yellow River region, and by 5000 BCE in the Korean peninsula and the Japanese islands.[7]

Domesticating pigs brought them within the human settlement and in closer contact with humans and other animals. But the early stages of the shift from nomadic to settled living were still a mixture of farming and hunter-gathering strategies. It is likely that pigs still had wide latitude in their travels and were left to fend for themselves in the fields, woods, garbage heaps and lavatories that surrounded these emerging settled villages rather than being confined to sties. In addition, wild food sources still contributed to the menu of these early communities. The Hemudu culture of the latter six millennia BCE, whose settlements were on the coastal plain east of present-day Shanghai, is a typical example. These people had domestic pigs and likely domestic water buffalo as well. The terrain surrounding the Hemudu towns was a mosaic of forests, lakes, meadows, streams, and marshes. In addition to early rice farming and the products of these domesticated animals, the trash heaps of this culture reveal substantial amounts of gathered foods such as acorns and water chestnuts, and a variety of hunted foods such as waterfowl. Bones from cranes, egrets, cormorants, and ducks were recovered from the excavated settlements. This mixed economy was a very successful strategy and supported growing human populations. It was likely common to a variety of cultures as they underwent the transformation from nomadic to settled.[8]

As we have seen, it is possible that prehistoric human populations were afflicted with an influenza strain that jumped directly from bird populations to humans. Since direct bird-to-human transmission is technically feasible, early influenza epidemics may have occurred in the Americas and Australia as well as in Eurasia and Africa. Migratory flyways link all the continents together, so an influenza virus could have been introduced by a variety of waterfowl, including gulls, terns, swans, geese, and ducks. But because the evolution of an enteric bird disease into a human respiratory disease requires a number of mutations, such a trans-species event had to be very rare. The introduction of the pig into human settlements made this disease transfer much more likely to occur, for, as we have seen, pig cellular anatomy may serve as a stepping stone for creating human transmissible influenza viruses.

The Neolithic Revolution had transformed human living patterns and the populations of settled humans were growing in numbers at tremendous rates, but these nodes of settled communities were still comparatively isolated from one another. Infectious diseases may have emerged and circulated within a larger circle of people, but the isolation of these populaces still imposed limits on the numbers of those at risk. However, the new food production system of farming did more than just compel people to become settled rather than nomadic. Farming led to storable surpluses that could be bartered directly with other settled communities or with semi-nomadic pastoralists. Surpluses could also be used to support artisans and merchants who could create different items of value. These products and new goods could be traded to others, sometimes across great distances, leading to sustained interactions with a number of populations. Storable food also supported kings and their attendant warrior classes, who may interact with their neighbors with less benign intent. Trade and conquest served to expand the linkages between growing populations. This resulted in a variety of exchanges, some of which were unintended.

The Confluence of Disease Pools

William McNeill in his landmark work *Plagues and Peoples* (1976) fully develops a concept he had first introduced in his massive *Rise of the West*, written a decade and a half earlier.[9] McNeill argues that growing populations sustained a variety of epidemic diseases. These afflictions had leapt from the domestic animal populations and continued to circulate in human communities that were sufficiently large to keep the chain of infection going. McNeill maintains that these populations, over time, had reached some sort of equilibrium with their pathogens. By this he means that although the illness was continually present, the people had adapted to the infectious agent to the point at which the pathogen infected a relatively small proportion of the total population, and was generally less virulent. As human populations grew and interaction with their neighbors accelerated, they eventually exchanged their infectious diseases so thoroughly that they could be considered part of a shared disease environment. He calls these groupings a "disease pool" in which the disparate infections of various smaller entities were combined into a variety of infectious elements that afflicted the entire mass of people. On the surface, this sounds like an innocuous process, but it actually must have been quite horrific. When populations came together (again through trade, travel, or conquest), infections that one group had reached equilibrium with—likely at great morbidity and mortality costs—were introduced into communities that had never been exposed to these pathogens and therefore had no immunity. The resulting linkages would spark epidemic spread among this newly connected population, since all were susceptible to the diseases to which they had not previously been exposed.

Just as was the case with smaller populations, these larger connected human settlements must have reached an equilibrium with their variety of pathogens. Over time, McNeill argues, three distinct large pools developed. One was centered in the Mediterranean, a second connected the poulations of the Indian subcontinent, and the third was located in East Asia. The creation of these individual pools must have occasioned a number of epidemics, but these went unrecorded. Beginning about 500 BCE, these large pools began to become interconnected and the process of creating a single, shared disease environment encompassing all of Eurasia and parts of Africa ensued. The evidence is fragmentary—and fierce debates have continued on the particulars (such as which came first to the Mediterranean, measles or smallpox?)—but both Chinese and Mediterranean history catalogue the devastating effects of epidemics of new diseases.

Turning our focus back to influenza, it is therefore not surprising that the first reports of what are suggested to be influenza epidemics appear during this time period of continent-wide interconnections. Many historians identify an outbreak of cough, fever, prostration, and pneumonia as described by Hippocrates in 412 BCE as the first recorded influenza epidemic. These historians point to the symptomology described by Hippocrates as well as the association of the outbreak with a sudden change of weather. In the northern and southern hemispheres, influenza seasonal onset follows the change to cooler, drier weather. Other early examples of presumed influenza outbreaks include Livy's account of a Roman encampment suddenly besieged by an epidemic in 212 BCE whose symptoms—as described by Livy—are considered indicative of influenza.[10]

It is impossible at this removed date to positively ascribe these and other infections to any one particular agent, especially as these writers lacked the tools to pinpoint and accurately diagnose diseases in the fashion we do today. But we can say that an environment had been created for the development of influenza epidemics. Humans were living in very close proximity to domesticated waterfowl, and, for that matter, wild waterfowl as well. They were also in close contact with swine, which have the capacity to serve as a bridge for the influenza virus, over which it may jump species and infect humans. In addition, a large, interconnected mass of humanity stretched the length of the Eurasian landmass and was linked with parts of Africa as well. The combination of these elements suggests that an influenza virus would have ample opportunity to be transferred to human populations, and once there, it would likely have undergone sustained serial transmission.

The virus could infiltrate the human population anywhere birds, pigs, and humans came together, so any region that had this shared disease pool could serve as the epicenter of an influenza epidemic.

However, the bird/pig/human interconnection was particularly tight in the region of China in which paddy rice production was prevalent. The Han Dynasty, which solidified its grip on the empire in 212 BCE, was known to have large numbers of domesticated ducks, and later writers have suggested that longtime experience made the Chinese "perfect masters in the management of ducks."[11] Over time, an efficient and integrated food production system developed in which ducks were placed on rice paddies. The ducks ate weeds and other pests while simultaneously fertilizing the growing rice plants. After the rice was harvested and the paddies drained, pigs were turned loose to scavenge the remaining stubble and unharvested grain while also fertilizing the soil. The combination of rice, waterfowl, and pigs served to recharge the nutrients in a plot of land and maximize its future food production. However, unbeknownst to these farmers, such a management method also ensured that pigs would regularly come into contact with the excretions of waterfowl. And, as we have seen, these often contain influenza virus strains.

While the conditions were ripe for the development of widespread influenza epidemics, we cannot speak authoritatively about them in this distant past. There are several reasons for history's silence concerning influenza epidemics, the primary one being the fragmentary nature of historical records. The vast majority of contemporary writings from the past have been lost, and those we have have often been passed down through copies and translations in which the text has been changed or distorted. It is possible that an authoritative catalogue of influenza outbreaks resides in an ancient Chinese chronicle or a Byzantine archive, but, if so, it has not yet been discovered and translated. A second reason for the inability to conclusively identify an influenza outbreak in the past is because the observations of an affliction by interested reporters necessarily relied upon a recitation of the physical manifestations of the illness. The sudden onset, fevers, prostration, coughs, aches, and in some cases the subsequent development of pneumonia can mark any number of illnesses. The fairly common assortment of complaints makes any definite diagnosis speculative at best. In addition, any proto-epidemiologist would have to want to pinpoint influenza against a backdrop of truly appalling diseases. In a world in which measles, smallpox, dysentery, diphtheria, plague, and any number of other afflictions run unfettered, it would be difficult for any chronicler to take note of, or even care about, an illness from which the vast number of people recover with few lasting ill effects.

Still, historians interested in tracing diseases have sought to uncover the tracks of past epidemics. In Europe during the Middle Ages, a number of widespread diseases that are described as having included fevers and coughs with a sudden onset have been asserted to be influenza. These records, however, are vague and incomplete. None the less, historians have intently studied these reports with an eye to retrospectively determining the cause of illnesses. Drawing upon these records, the famed August Hirsch identified an influenza epidemic in the year 1173. Examining testimonies from later centuries, Hirsch believed he had detected the intermittent reappearance of influenza outbreaks, and although the evidence he reviewed was insufficient to provide a measure of the geographic scope of the illness, he was confident in identifying the limited afflictions as influenza epidemics. These epidemics remained somewhat localized, however. Hirsch maintained that it was not until 1510 that the first "truly pandemic forms" of the illness appeared.[12] Relying primarily upon these fragmentary reports, Hirsch drafted a chart of influenza outbreaks from 1173 to 1874. Although Hirsch was confident in his identifications, it is clear that these retrospective diagnoses are highly speculative, especially the ones most removed from Hirsch's time of writing in the mid-nineteenth century.

One can begin to have more confidence in more recent retrospective influenza outbreak declarations. The number of testimonials and eyewitness accounts available for consultation began to increase, and the descriptive skills of these observers also improved. A revitalized focus on the human body, and an attempt to clearly discuss aspects of it, began to become prevalent in select circles in a variety of societies. It is thus no surprise that Hirsch's first identification of true pandemic spread dates from an illness outbreak in 1510, as the Renaissance took hold in Europe. Indeed, it was those paragons of the Renaissance, the Italians, who provided us with the name "influenza," a term they used to identify an outbreak in sickness in 1504 that they chalked to the "influence" (*influenza*) of the stars.[13]

Although we cannot say for sure, the illnesses these sixteenth-and seventeenth-century observers described certainly sound like an influenza epidemic. Theophilus Thompson, writing in his *Annals of Influenza* (1852), quoted a Dr Thomas Short, who reported that the disease outbreak of 1510 "attacked at once, and raged all over Europe [sic], not missing a family, and scarce a person." The symptoms described were "a grievous pain of the head," difficulty breathing, loss of strength, and a "terrible tarring cough." These complaints were followed by chills and fevers, but few died, except some children. In similar fashion, the description of a sudden illness in 1658 featured an identical litany of ailments—fever, cough, pains, prostration—and the compiler, a Dr Willis, noted that the whole sweep of affliction began and ended in "one month's space." The continuity of symptoms from the various epidemics he surveyed prompted Thompson to remark: "one of the most remarkable circumstances impressed on our notice, is the great similarity of symptoms presented by the disease in its different visitations, notwithstanding every diversity of season and place."[14] It should be noted that the reports consulted by Thompson and other writers were largely derived from direct observation, generally from the patients these physicians had sought to cure from this sudden illness. These early reports generally made little effort to collect data systematically or to assess the scale and scope of the outbreaks.

The chronicle of complaints and a rudimentary tracing of the spread of the (presumed) influenza outbreaks were at first very useful to understanding the true impact of influenza epidemics, but the evidence collected was usually episodic. It would take the famed father of clinical medicine and epidemiology, Thomas Sydenham, to move beyond the mere tracing of the affliction. His account of the "epidemic coughs of the year 1675" tracks the sudden widespread onset and sharp decline of the illness in London. Sydenham charts the appearance of the new illness in late October, with the widespread sickness declining towards its disappearance by the end of November. What separates his account from those of previous chroniclers is that Sydenham tried to assess the impact of the epidemic beyond mere anecdotal assessments. Instead, he sought to account for those whom the epidemic killed. Generally, previous observers had merely noted that almost all of their patients recovered, apart from some who were very old or very young, and that even these deaths were small in number. Sydenham's study of mortality in the city of London unmistakably revealed a climb in the death rates that trailed the onset of the illness and reached a peak in the middle of November. This mirrored his observations of sickness rates associated with the new affliction. The peak rate of mortality in mid-November was double the rate in the weeks prior to and after the mystery affliction raged. His careful study revealed that these "epidemic coughs" were no mere nuisance, but were deadly. In the new age of Enlightenment, data was beginning to replace impressions and conjectures.[15]

As the Enlightenment ideas spread and "scientific" approaches began to be adopted in a variety of places, the rapid spread of influenza epidemics became much more richly documented. Whereas the epidemic of 1580 had been linked across a wide sweep of territory from Asia to Africa because a variety of interested observers in different places took note of the local appearance of the illness, the epidemic of 1781 was systematically charted on its continent-wide jaunt. The illness was first detected among the members of the British Army stationed in India in November. China was identified as similarly undergoing the infection, with cases identified in Siberia and Russia in December, quickly followed by German outbreaks in February. April brought the affliction to the Scandinavian countries, trailed by England and Scotland in May and France and Italy in June, the disease finally concluding its Eurasian march in Spain in August. The nature of the causative agent remained a mystery, but the careful tracking of the epidemic's journey suggested that the illness was contagious.[16]

Thomas Sydenham had taken the first halting steps to measure the full impact of an influenza epidemic in the latter seventeenth century, but it would take more refined measurements to gauge influenza's real mortality toll in a community. William Farr, famed epidemiologist of nineteenth-century Britain, was the man who designed a truly effective model to assess the mortality impact of illnesses on a society. Farr wedded the power of the new science of statistics with the bureaucratic power of the modern state,[17] parlaying his position as "compiler of abstracts" in the newly created Registration General Office to systematize the collection of the nation's vital statistics.

These—the records of births and deaths—provided a valuable snapshot of the health of the public. In addition to requiring every county and town to submit these figures, Farr pushed for a standardized diagnosis of the cause of death. From this rich collection of data, Farr developed a statistical form of measurement he called "life tables." These life tables generated an expected range in the number of deaths in various places that reported their yearly death rates—ranging in size from individual towns to counties in England and Wales. Whenever the number of deaths exceeded this threshold, Farr knew that something unusual had been going on in that region. Often, he determined that the increase was due to the circulation of some ailment. Farr sought to craft "laws of mortality" to predict the number of expected deaths, and he endeavored to develop a numerical value for the economic costs of the loss of human life. Farr discovered that the mortality rates differed significantly between rural and town populations. This he ascribed to the increase of epidemic diseases, which, he noted, spread more efficiently in denser populations. Farr's assessment of the "value" of human life sketched out the costs to the nation of a life cut short by epidemic diseases, providing a crude measurement of the toll of influenza outbreaks.

As it turns out, the notion of excess deaths—deaths above the expected rate of mortality—is crucial for appreciating how dangerous an influenza outbreak can be. Determining why someone had died was difficult for these proto-pathologists and so the cause of death was often listed somewhat generally as fever, pneumonia, old age, or any number of general causes. Identifying why any one individual died was difficult, at least beyond the immediately obvious. This difficulty in diagnosis is particularly true of an influenza illness, since the symptoms of a case of influenza are also common to other diseases and because the death of the patient may trail the initial infection by a lengthy time period. It is for this reason that Farr's life tables and excess mortality model was so crucial for assessing the impact of influenza epidemics. What Farr's statistics revealed was that mortality across a variety of categories rose in lock step with the epidemic wave pattern of an influenza epidemic. Farr recorded the listings of "pneumonia," "pleurisy," or even "old age" that physicians around Great Britain sent to his office, and plotted these deaths against his life tables. What he discovered was that mortality increased and declined across a range of death categories in a ghostly echo of the arrival of an influenza epidemic. The conclusion was unmistakable: the influenza epidemic had killed a number of people, more than just a simple reading of the causes of deaths reported by physicians would indicate. These extra deaths, across a variety of categories, were due to an influenza infection in some way, and this impact could be measured statistically.

The experience of London in the influenza pandemic year of 1847–8 illustrates both the impact of influenza and the power of Farr's new tool.[18] As Charles Creighton described it, this was the first great epidemic of influenza since the new system of vital statistics registration. The epidemic caused an excess of 5,000 deaths beyond what was expected during the six weeks it raged in London, but only a quarter of these were officially ascribed to influenza. The rest had been chalked up to pneumonia, asthma, consumption, or some other ailment. The tables revealed that at the height of the epidemic, deaths in childhood increased by 83 per cent; in manhood 104 per cent; and in old age 247 per cent. At the same time, deaths in young people between the ages of 10 and 25 barely increased at all. Here we can see the general pattern of who is most threatened by an influenza epidemic, what we today call the "high-risk" population. Statistics that only collected deaths labeled as due to influenza indicated only a fraction of the true mortality impact. It also bears noting that methodology not too different from Farr's lies at the core of assessing the impact of influenza in the present day. Determining the number of deaths and illnesses above an expected threshold is still a key measurement in declaring a pandemic and assessing its true costs.

Farr's statistical tool and the careful tracking of the sudden appearance and steep rise of illness associated with the epidemic revealed the outlines of the impact of influenza for the first time. But these figures and assessments did not solve the issue of the cause of the sickness. In fact, charting the near-simultaneous eruption of the infection in such disparate locations may have helped to cloud the proper identification of the virus's contagious nature. Because the epidemics appeared so widely—and, as many noted, seem to trail a sudden change in the weather to colder temperatures—physicians sought some climatological explanation for the affliction. Others took note of how truly widespread a pandemic was and ascribed it to cosmological causes: for example, that the illness rained down from space. Some still clung to a "malarious vapor" as the method of disease transmission while others asserted that the illness was the result of the spread of some sort of contagion. These various explanations of the onset of influenza epidemics were part of a larger argument over whether health and sickness were the result of some internal complaint manifesting in disease (humoral imbalance), environmental causes (bad air), or some sort of contagious transfer. The upshot of all this debate was that no clearly accepted element was identified as the source of influenza epidemics for almost the entire nineteenth century. As we shall see, a late nineteenth-century pandemic was chalked up to an infectious agent, which generated widespread agreement. This identification turned out to be incorrect, but the misidentification proved to be very valuable in detecting the true cause of influenza pandemics and unraveling the mystery of influenza's origin and course of infection.

Influenza was likely introduced into the human population as the result of a revolutionary change in food production and living patterns. Throughout the age of Enlightenment and into what historians terms the Modern Era, another revolution was brewing. Changes in the ways human lived and worked were occurring as the result of expanding industrialization and concomitant urbanization. The rise of powerful empires and the expanding reach of their colonial ambitions served to bring all regions of the globe into closer contact. The accelerating pace of modernization ratcheted up the speed with which far-flung peoples were interconnected and the voracious appetite for trade and plunder greatly increased the volume of this contact. Intertwined with this process (historical demographers still debate the causes of this phenomena) was soaring population growth. The end result of this revolution, a transformation in which you and I are enmeshed, is a tightly interconnected global world in which an increasing majority of people live closely together in urban regions and are connected to every other person on the planet by rapid and voluminous transportation networks. In this new world, every person truly is our neighbor.

The aftermaths of revolutions are times rich with possibilities, both good and bad. It is well beyond the scope of this book to assess the good and bad of the Modern Era. Limiting ourselves to the topic of influenza, one can see that the changes have created a tightly linked disease environment in which a novel strain of influenza can quickly become a true *pan demos*. The twentieth century generated a world in which influenza viruses could speedily circulate with deadly effect. This large single-disease environment facilitated the rapid spread of infectious organisms, including influenza. As Farr showed, influenza epidemics are not the innocuous disease events that had generally been assumed, and, as we shall see, one strain raged with a power never before (or since) observed, demonstrating how truly devastating the virus can be. But a world which all jointly inhabit is one that encourages shared attempts for mutual protection. The twentieth century marks both the threats of a single-disease environment and the potential promise of international cooperation.

THREE

Misidentifications and False Starts

The nineteenth century marks a transition point between two different worlds. At the outset of the nineteenth century, the local still held pride of place. By the close of the century the international or global began to rise to ascendancy. The early 1800s was an era that was still driven by rural and agrarian concerns and travel was difficult and time-consuming. By the late 1800s, many societies were transitioning into being urban and interconnected with rapid advances in communications and technology—notably applied to transportation—which made it easier and quicker to travel and transport materials across vast distances. These changes were not merely a function of events in the nineteenth century, of course. After all, it was Columbus and Magellan and those that sailed in their wake who had, as Alfred Crosby so memorably termed it, re-knit the seams of Pangaea.[1] And the long-distance transportation routes and the urban areas that coalesced at various nodes of the vast Silk Road network predate the system that arose in the late nineteenth century by many hundreds of years.[2] But it is also unmistakable that the dawning of the twentieth century heralded a radically transformed world; one more truly global than ever before, which increasingly moved to faster rhythms.

It is often difficult for people living in a particular time period to identify the transitions that are reshaping their lives, especially those that spin out over an extended period of time. Future historians may label an age as the "beginning of industrialization," or the "steam power revolution," which carries with it the implicit notion that one way of living is abandoned for another. But the experiences of lived history are not so readily bifurcated. The transformations unfolding may occur rapidly in one place, slowly and subtly in another, and not at all in a third. The same holds true for ideas as well as technologies. New concepts compete with old explanations and there is often a period of overlap between the adoption of a new approach and the abandonment of the old. Often, it takes some sort of crisis to bring into high relief the changes of the world, and this is also true in the world of ideas. During challenging events old practices or explanations are put to the test as society is shaken. In such trying periods, new patterns of society organization or new intellectual concepts may emerge to replace older systems that have faltered under the pressures of events. These tests of the status quo can take many forms: economic failure, natural disaster, war, and disease being some of the more prominent. A pandemic can serve as a serious threat to a society, either because the current societal organization is unable to handle the problems the epidemic induces, or because the illness has shattered the conventional understanding of sickness and health. Crisis events illuminate changing societies.

Pandemics in the nineteenth century illustrated that urban areas were growing in both size and importance in the nation-state and that these urban areas were clearly tied to one another by speedy transportation networks. A new, interconnected world was emerging in which the large and growing cities of states could be more linked, and have more in common with one another, than with other regions in the same state. The role of urban areas and transportation—notably shipping—presented a series of difficult challenges for coming to grips with this changing world. Shipping networks were important for bringing the benefits of trade and travel, but they could also inadvertently serve as the deliverers of pestilence and sickness between states. The linkages between urban areas from a variety of states meant that safeguarding against disease was not just a local, but more and more an international concern.

The close relations between states, and between the homeland and the overseas empires many of these states had come to command, brought exposure to a catalogue of new afflictions. The increasing contact with an array of novel diseases, and the apparent differing levels of susceptibility to infections by various populations, began to challenge accepted notions of medical treatment. Nineteenth-century ideas about how to maintain good health still drew upon the notions of humoral balance first proposed by the ancient Greeks. Epidemics did not fit easily into this framework of medicine and so competing explanations for them were proposed, leading to fierce debates in the scientific and medical fields. These discussions became more than just academic arguments between physicians, as a number of afflictions began to speed along the transportation routes of the nineteenth-century world. These conflicting debates as to the origin, movement, and spread of diseases came to a head in the Russian flu of 1889. Close study of the pandemic, and the identification of a presumed entity responsible for the disease, conclusively demolished the humoral notion of health and helped to cement the new theory of disease onset. The succeeding influenza pandemic, the calamitous Spanish flu, would challenge the accepted explanation for influenza and so help to focus study on a new disease-causing organism.

Although influenza served as an important test for a new theory of the outbreak of diseases, this new model was developed as the result of another series of pandemics: cholera. Cholera shook the foundation of science and public health and challenged the prevailing notions of medical practitioners. It brought into the open the sometimes intertwined debates between contagion, environment, and individual predilection in the onset of illness, and it was this feared and dreaded disease that played a key role in the development of the germ theory.

Nineteenth-century Cholera Pandemics

Cholera was the first great nineteenth-century pandemic that served to highlight the interactions between urbanization, transportation, and international connections.[3] Breaking the bonds of its traditional epicenter in the Ganges river watershed, the affliction radiated out in a series of pandemics that swept the globe over the course of the nineteenth and into the twentieth century. Cholera presented a number of challenges to nineteenth-century governments and scientific thinkers. First, it was inextricably tied up with colonization and imperialism. India was the jewel in the crown of the nineteenth-century British Empire, and imperial policies helped both to transfer the infection and to prevent the imposition of quarantine and embargo practices that would have minimized the pandemic's spread. Second, the illness clearly illustrated the increasing importance of urbanization, industrialization, and transportation. Finally, the epidemic presented a difficult puzzle for medical understanding of disease causation, since it could be marshaled to support any number of different interpretations for sickness. Eventually, the source of the disease would be defined and this identification would be used to bolster similar investigations into the origin of diseases such as influenza.

In 1817, a major cholera epidemic erupted in India, where it had been a scourge for centuries. Unlike in previous cholera outbreaks, however, the illness began to be carried beyond the subcontinent. The affliction was brought as far east as China (which likely had endured cholera outbreaks previously, but certainly not as regularly as in India) and as far north as the Russian steppe. In 1820, as the pandemic continued, British soldiers who were shifted from India to the Persian Gulf inadvertently brought cholera with them, infecting the population of the eastern Mediterranean. The disease circulated for a period of time, but how many were infected and died was not accurately recorded. Eventually, the spread of the illness sputtered out. Although the epidemic was an abortive pandemic, the fact that the illness had traveled out of the subcontinent was a harbinger of what was to come.

In 1832, a larger epidemic emerged and circulated out from India. In addition to infecting Russia and China again, the illness appeared in European, North, Central, and South American cities, making this wave of infections a true pandemic. Subsequently, large pandemics began in the years of 1848 and 1866, with lesser outbreaks spreading in several other years. It was characteristic of the transmission pattern of the illness that it would appear first in cities, generally port cities linked to trade. Outbreaks could be tied at first to the waterways that linked a region and in later years to the expanding railroad systems that crisscrossed states. The slow-moving spread of the infection served to heighten the fear and dread of its (well-publicized) arrival.[4]

Cholera challenged traditional assumptions of health and disease. We now know that cholera is an illness caused by a microorganism known as a vibrio. The disease is generally transferred by fecal contamination of water and food sources. In order to infect the body, the vibrio must first survive the harsh acidic environment of the human stomach. Once past the stomach, the organism colonizes the small bowel, attaching itself to the lining of the intestine. To facilitate its continued transference, the organism releases a powerful toxin to which the body responds by trying to flush it out. The resulting massive evacuation of fluids leads to dehydration and the loss of vital salts and minerals necessary for the function of the body, as well as broadcasting the vibrio into the environment where it can infect others. Left untreated (and in the nineteenth century there was no understanding of the proper treatment for this infection), the toxic nineteenth-century version of cholera led to rapid and painful death in approximately half of those infected.[5] Aspects of the epidemic were malleable enough to support a variety of disease outbreak explanations, which competed for support in the nineteenth-century intellectual world.

Contagionists argued that "Asiatic cholera" (as the disease was named) was caused by the transfer of some contagious material or entity, although what that agent was remained unclear. Like the two recognized contagious illnesses—plague and smallpox—the appearance of one infected person was soon followed by a number of others. And one could track the course of the pandemic as it spread from city to city—something the newspapers of the day blared from their headlines. But the traditional tactic of using quarantine to keep contagious illnesses at bay was ineffectual against cholera: holding ships from infected ports in harbor did not prevent the appearance of the disease. Nor could contagionists explain why the affliction struck so suddenly in communities and among people who clearly had not had any interaction with the initial infections.

Anti-contagionists (broadly grouped) argued that environmental conditions were responsible for the eruption of disease. This environment-based interpretation argued that dirt and filth created odors that generated miasmas, which prompted pestilence. This surprisingly durable and flexible construct of disease causation was fundamentally a medieval explanation. The environment-based disease concept evolved into a movement of medical and scientific personnel concerned with protecting the public's health, called "sanitarians." Sanitarians rose to the fore in the mid- to late nineteenth century and called for a variety of techniques—including regular emptying of privy vaults, the construction of efficient sewerage systems, the draining of stagnant bodies of water, and regular trash collection and street-cleaning—to remove the sources of noxious and what they deemed dangerous fume-producing material from towns and cities.[6] The events of a cholera epidemic bolstered this environmental explanation.

In the case of cholera, anti-contagionists could point to the fact that the disease usually first appeared in the most squalid sections of town—coincidentally, often the areas along the waterfront in port cities—and that the illness raged most devastatingly where the population lived in the most fetid circumstances. However, such an explanation failed to account for the obvious serial transfer of the infection from city to city. Some anti-contagionists modified their positions by stating that the contagious element required interaction with the environment before it could produce disease: this was the so-called "XYZ" concept most vigorously promoted by Munich scientist Max von Pettenkofer, who argued that contagious element "X" needed to interact with the right soil or water "Y" to generate the disease cholera "Z." Without these linked connections, the disease could not appear.[7]

A third explanation blamed the individual for their sickness either through sinful behavior or through a weakened physical state due to dissolute living. The "judgments of a vengeful God" thesis generally held sway in describing epidemics early in the century, while the more "scientific" explanation of a weakened constitution was more prevalent towards its close. Both centered on the fact that cholera usually first broke out in the impoverished or immigrant sections of town where the ungodly or intemperate congregated. In the nineteenth century, of course, poverty was seen as the result of personal or moral failings.[8]

Again, the reality of cholera epidemics undercut such explanations. While the outbreak may strike its heaviest blow on the poorest section of town, fatalities occured almost everywhere, even among the wealthiest and most upright members of a community. While whispers of secret moral failings may suffice for a few cases, over time the "judgment of God solution" proved inadequate to account for the real mortality pattern of the epidemic. Eventually, the notion developed that the poverty-ridden were a threat to the "better sorts."

In the United States this often took the form of blaming immigrant groups and calling for their exclusion or confinement.[9]

The solution for this vexing disease-causation problem appeared in the 1860s, although it was not immediately embraced.[10] The germ theory developed by Louis Pasteur (as the new model came to be called), was initially just another explanation for disease outbreak. Vigorously promoted and enlarged by the work of Robert Koch, this argument received a strong fillip with Koch's discovery of the organisms responsible for anthrax, tuberculosis, and cholera. Pasteur, Koch, and their followers forcefully argued that the eruption of disease could be traced to microscopic organisms that invaded and replicated in the human host, prompting the infections. The "germ" was a discrete entity and not an amorphous vapor. The germ theory explanation solved the confounding riddle of cholera epidemics. Cholera was a contagious disease prompted by the proliferation of the "comma-shaped" bacteria known as vibrios. It was spread by infected water sources—validating the linkage proposed by John Snow in his classic epidemiological studies of 1854—and the victim was infected by coming into contact with or ingesting cholera-infested water (the chance factor of its surviving the acidic environment of the human stomach was not yet appreciated).[11] Poorer areas suffered disproportionately high rates of infection because the disastrous state of their water procurement and sewerage processes meant that they were more likely to come in contact with the organisms. But others could be struck too; the key was contact with the vibrio. Therefore, the way to protect a population against cholera was to isolate the stricken and to ensure fresh water supplies and the efficient removal of wastes.

Throughout the 1880s, the germ theory and opposition to it roiled scientific and medical circles. Although it gained adherents, the new theory had by no means overcome all objections and the debate over disease causation was not settled. To a large extent, influenza had not been marked by these disputes. No microscopic entity had been identified as the root of the sickness, which itself had a very low profile. Influenza activity had entered a quiet period since the epidemic years of 1847–8 that had drawn Farr's attention. Certainly, the disease continued to circulate, but not in the dramatic pandemic wave pattern that was so obvious and disruptive. Although it is impossible to be positive at this removed date, it is all but certain that influenza appeared seasonally, infecting and killing some percentage of the population. But such illnesses faded into the backdrop of a world where fevers, coughs, pneumonias, and a slew of other complaints circulated unfettered. Influenza's quiet period was soon to end, however. Its eruption into a pandemic starkly illustrated the speed and interconnectedness that new transportation technologies created, and provided a venue in which to test the new germ theory's validity.

The Russian Flu of 1889

In the second half of May 1889, reports began to filter west that a great number of people in and around Bukhara had been stricken with an affliction of fever, prostration, and head and body aches.[12] At the time, these reports attracted little notice. Bukhara, a Central Asian city (currently part of Uzbekistan), had a long history as an important site on the Silk Road, the vast and ancient interlinked trading network that connected China to the Mediterranean. By July, half the Europeans dwelling in the town had suffered the affliction, but throughout the rest of the summer few other noteworthy outbreaks were reported. Very quickly, however, influenza would force people to pay attention to it again.

In early November 1889, it was reported that the citizens of St Petersburg were suffering a mystery illness that left a great number prostrate with a mixture of symptoms including coughs, fatigue, respiratory problems, and head and body aches: the usual hallmarks of influenza. Subsequently, it was determined that the illness had appeared in late September and the number of those stricken rose steeply throughout the month of October. Readers of the news reports did not have long to ponder over the identity of the mystery affliction, for in very short order the sickness was upon them. Assuming early October as its starting date, within six to eight weeks the epidemic had swept over the whole of Europe and penetrated North America. Within two months, it had reached the Cape of Africa. Within three months, South America was enveloped and by month four, it was circulating in India. In the fifth month, it could be found in Australia and New Zealand. In just over half a year, the illness had become a pandemic, touching virtually every population center on the planet.[13]

Physicians and public health spokesmen initially attempted to downplay the sweeping infection. It was, after all, only the "grippe," a minor annoyance. Initially, the epidemic outbreak attracted little attention in the European press. For example, the first accounts of the illness in Russia appeared in the *nouvelles étranges* of the French papers. But, as the affliction raced across the continent and the scale of the pandemic began to become clearer, the stories were moved to the more prominent general news section.[14] Even as the pandemic arrived and the number of people infected began to soar, physicians maintained that there was little to be concerned about. In France, medical and public health personnel clung to their initial promises that the illness was "neither deadly nor dangerous," while in Barcelona, prominent physicians pronounced that the disease would not be severe so long as patients "strictly followed [their] physician's advice."[15]

In truth, such promises were empty rhetoric. Increasing numbers of people were dying and much greater numbers were laid low by the illness. Despite a laundry list of attempted procedures, physicians had little to offer in terms of protecting or curing their patients. As is often the case, a number of self-proclaimed healers peddled an amazing array of nostrums, elixirs, and potions to protect against the "grippe."[16]

If physicians were powerless to treat patients incapacitated with the flu, the unfolding pandemic provided an opportunity to put the validity of the new germ theory to the test. Promoters and detractors of the new disease explanation studied the unfolding epidemic, generating a rich variety of observations in a number of places.[17] In Great Britain, for example, the Medical Department of the Local Government Board placed a letter in the 11 January 1890 edition of the *British Medical Journal* inviting its readers to help map the appearance, transmission pattern, symptomology, and duration of the illness. This solicitation, and the response by a number of interested medical officials, marked the first time an influenza epidemic (and perhaps the first epidemic at all) was tracked in real time.[18] The questions about the nature of the disease of influenza were coming into focus.

Influenza was another disease that had been cited by all sides of the disease-causation debate. Its appearance at different localities in different times suggested the serial transfer aspects of a contagious disease. But its sudden, widespread outbreaks were pointed to as proof that the source of the infection was environmental or meteorological. The last pandemic year of 1847–8 predated Pasteur's development of the germ theory. The close tracking of the Russian flu began to solve the questions of disease onset and transfer in the case of influenza. As the reports generated by observers began to be compiled, a pattern emerged. First, the illness was clearly caused by some sort of contagion. The epidemic moved from place to place (generally from east to west) and a smattering of cases appeared before the steep epidemic wave of victims ensued. Second, the movement of the disease was facilitated by transportation systems and was related to the size of the population. The first cases in a region appeared in stops along rail lines and canals, and from these locations spread to the surrounding towns and countryside. When bodies of water interrupted land travel, the ports connected to sea travel served as the initial sites of influenza infection, which was subsequently dispersed by the transportation network. Third, the illness prompted a very steep rise and rapid drop-off in infections that when graphed produced a distinct wave-like pattern. And, as the years 1891, 1892, and 1893 would demonstrate, the affliction could reappear as a wave months later. Finally, although a vast number of people recovered, mortality increased in a number of categories, with the aged or those with some pre-existing health problem (for example, tuberculosis, which was the major killer disease in the nineteenth-century world) providing the greatest percentage of the fatalities. The sum total of these observations was that influenza epidemics were very disruptive of society. A very conservative estimate for first-wave mortality (1889–90) in Europe posited that the number of deaths had been between 270,000 and 360,000.[19]

In addition to being inspired to track the unfolding pandemic, adherents of the germ theory sought to identify its causative organism. And the illness obliged these researchers with a large number of patients to examine as the sickness returned in subsequent waves up through the year 1893. Scientists peered through their microscopes at a number of different samples extracted from the sick, as well as material from post-mortem cases. Although the symptomology of coughs and respiratory distress prompted them to focus on the nose, throat, and lungs as the harboring place of the contagious element, no portion of the anatomy was overlooked in the search for the germ that caused influenza. A variety of cultures were tried and a spectrum of stains applied in the hope of capturing and cultivating the unknown element. Closely examining the sputum of infected patients, especially patients who subsequently developed pneumonia, generated a variety of potential culprits. Although some of the bacilli observed had their champions as the germ responsible for influenza, none could be consistently found in all the sick, and often these same microbes could be found in those who were not ill with the disease. Influenza remained puzzling.

In 1892, German bacteriologist Richard Pfeiffer reported that he had been able to recover a bacillus found in the sputum of infected patients which he believed to be the agent responsible for the illness known as influenza. The microbe, which became popularly known as "Pfeiffer's bacillus," was difficult to culture, but Pfeiffer asserted that it could be found in active influenza cases, sometimes almost in pure culture, meaning that colonies of only that bacteria could be recovered from patients. Generally, humans harbor a number of bacteria types; a pure culture implies that these bacilli are solely responsible for the illness. This difficulty in cultivating the microbe went a long way to satisfying critics, who were unable to find "Pfeiffer's bacillus" in their active influenza cases. Pfeiffer's stature in the field of bacteriology— he was a well-respected assistant to Robert Koch—and the compelling evidence he presented proved decisive. Within a few years, it was generally agreed in the medical field that "Pfeiffer's bacillus" was the "germ" responsible for influenza. When researchers suspected influenza, they looked for the distinctive microbe culpable for the illness. This conviction that influenza outbreaks had a bacterial origin was to hold sway until it was tested during the next great pandemic.[20]

As the final waves of the Russian flu pandemic petered out, influenza again receded into the background. By the early twentieth century, the germ theory was generally accepted in the medical field and the idea of a specific germ being the cause of the illness was beginning to filter into the consciousness of the general public.[21] Those who rejected the germ theory grew fewer in numbers over the years and were generally ignored or silenced. When influenza cases were suspected, it was assumed that "Pfeiffer's bacillus" was the cause. If a researcher was unable to detect the microbe, it must be due to the difficulty of cultivating these fickle bacteria.

Influenza would put specialists' scientific and medical knowledge and skills to the test in the early twentieth century, a test that they were doomed to fail. Spanish flu upended all the accepted thinking on causation, lethality, and those who stood at most risk during an influenza pandemic, and disillusioned researchers who had believed that mastery of the illness was imminent.

Spanish Flu

The worst influenza pandemic in recorded human history started modestly.[22] In early March 1918, the infirmary at a military encampment in Kansas was flooded with soldiers complaining of fevers, respiratory distress, body-and-head aches, and exhaustion: in other words, the flu.[23] This illness rapidly spread to other military camps in the United States and from there to civilian populations. The rapidly escalating epidemic traversed the Atlantic, and by May it was laying out soldiers on both sides of the trenches. Soon the civilian populations of all the states of Europe were infected. In the distinctive wave pattern familiar to those medical experts who had studied the Russian flu, the sharp rise and sudden drop-off of those afflicted with the illness could be traced globally—this time traveling from west to east —and again the role of transportation linkages appeared in high relief.

But this spring and summer epidemic, now officially a pandemic as it continued to unroll around the planet, was only a minor concern in 1918. A bout of the flu was, of course, inconvenient, but the illness's spring nickname of "three-day fever" hints at the relative unimportance the wave of contagion was thought to have. All eyes were locked on the unfolding events of the Western Front in the early summer of 1918. The hideous Passion play of the First World War was reaching its denouement as Germany used all its strength in a last desperate gamble to capture Paris and conclude the war in its favor. What interest and attention could an epidemic of the flu command while the outcome of the war hung in the balance? But nature was soon to demonstrate again that it had its own plans, regardless of those of humankind.

By the early twentieth century, the planet was firmly encircled in a vast transportation network. To an extent even greater than during the Russian flu of the late nineteenth century, the cities of the globe were part of one entangled human web. Prior to the onset of war, seaways and railroads hummed with trade and travelers. Although the outbreak of war had cut some connections, others rapidly increased and expanded. The terrible engines of this cataclysmic war required massive and accelerating inputs of men and material. The colonial reach of the Great Powers drew heavier and heavier on their overseas resources. And for those that they could not command, the combatants served as a giant market for the produce of other states, because the voracious appetite of the struggle required a staggering variety of things—such as wheat, horses, rubber, munitions, and most of all men, among a host of other needs—in great quantities.

The Great War had created an unusual confluence of events. Many millions of men and women were crowded together in substandard conditions as the conflict brought hordes of soldiers and their support personnel in close proximity on either side of the contested front. These crowded masses were firmly linked to transportation systems that flowed to every corner of the globe. The stressed population comprised a vast tinderbox for epidemic outbreaks—especially respiratory epidemics. None were more aware of this than the medical services of the various states; and thus far, the spring and summer flu cases aside, they had largely been successful in keeping at bay those epidemics that had ravaged previous armies in human history. This was about to change.[24]

Close study of the Russian flu revealed that the infection had traveled in a great wave pattern that swept through a district rapidly. The whole affair of appearance, steep rise, and sudden decline could unfold in a matter of weeks. In the pandemic of 1889, Russian flu had reappeared months later in a region as a second and still later as a third wave of cases, mostly infecting those not attacked in the previous waves. In some of these regions, Sheffield, England, for example, this second wave was even sharper and more deadly than the first.[25] But this more deadly second wave was the exception rather than the rule. In the Spanish flu the second wave was dramatically more deadly; so deadly, in fact, that the second wave is referred to as the "killer wave," since it was responsible for the overwhelming majority of deaths in the pandemic. It is the unusual scope—and, as we shall see, the unusual pattern—of its victims that has continued to puzzle and fascinate influenza researchers.

Spanish Flu's Second Wave

In late August 1918 the second wave erupted almost simultaneously in three diverse locations: Brest, France; Boston, Massachusetts; and Freetown, Sierra Leone. Although geographically dispersed, all three were port cities and firmly connected to the great transportation networks that thrummed with life during the Great War. Boston was a major exit point for the men and material streaming from the United States to support the Allies in Europe. Brest was one of the major receiving ports for this deluge of equipment and manpower.[26] Sierra Leone was one of the largest coaling stations that linked the Indian Ocean and the Pacific with the North Atlantic and Europe. Although in some ways these port cities were merely repeating the role port cities have often played in the past—as the entry point for epidemic diseases—this second wave of Spanish flu was both qualitatively and quantitatively different from any previous epidemic, except perhaps for the fearsome and deadly Black Death.

This second wave of Spanish flu was explosively infectious, especially in confined places like ships or prisons.[27] Ultimately, it was estimated that nearly three out of ten people on the planet were stricken by the infection in only a few months.[28] In addition, a significantly higher percentage of the infected became seriously ill. The onset was dramatic and the symptomology seemed direr, with higher fevers and a greater intensity of prostration and head and body aches. Most ominously, there were many more victims whose illness progressed into pneumonia. The combination of the greater number of victims and the greater percentage of severe cases rapidly overwhelmed hospital and medical infrastructures.

A classic and well-documented account of Spanish flu's second wave comes from Camp Devens, an army base near Boston that was scandalously overcrowded.[29] Approximately 5,000 of the camp's

45,000 troops were quartered in tents, and all the barracks were densely overpopulated. Such cramped housing arrangements predominated at military encampments as the United States strained to train and transship soldiers of the American Expeditionary Force (AEF) for service in the trenches of France. Devens, too, had felt the impact of the spring wave of "knock-me-down-fever," but neither the camp nor its medical department was prepared for the onslaught of the second wave of Spanish flu.

On 28 August 1918, sailors working on the Commonwealth Pier in Boston began to be laid up with what appeared to be influenza. By 7 September, the illness had made its appearance at Camp Devens (about 30 miles outside the city). On that day 96 soldiers were admitted to the hospital. On 10 September, an additional 142 soldiers were taken in. By 15 September, new patient admissions had risen to 705. New hospital admissions reached a peak over the next three days as 1,189, 1,056, and 1,176 stricken members of the camp were hospitalized. Such an accounting only documents those admitted to the hospital (with records) and fails to number those who were treated in barracks or who were not brought to the overcrowded hospital at all. Following this peak, the rate of new admissions began a steep decline, dropping to 684 new patients on 21 September. The number of those newly stricken would continue to steadily decline over the ensuing few weeks.

Here, then, we can clearly see the wave pattern of pandemic influenza. From no hospitalized influenza cases in the first week of September, admittances to the hospital soared to a peak by week three, and from there began a steady decline over the ensuing weeks, dropping below 100 new cases by early October. Ultimately, one-third of the men at Devens were diagnosed as having contracted influenza in September and October. And many of these men and women afflicted —for the medical staff at the camp were equally susceptible—were far sicker than had been previously seen with flu outbreaks. In a ghostly echo that trailed the hospital admission rate for the camp, the number of cases diagnosed with pneumonia traced an epidemic curve in miniature.[30] In the first week of the disastrous second wave's appearance at the camp, 57 soldiers were identified as having pneumonia. By the week ending 20 September, that number had climbed to 207, reaching a stunning peak of 2,023 new admissions for the week ending 27 September. In a world without antibiotics, there was a good chance that pneumonia meant death. Ultimately, 787 of the estimated 17,000 infected people at Camp Devens died, most due to complications from pneumonia.

Mortality rates also reveal the wave pattern in pandemic influenza. In Philadelphia, a city of approximately 1.7 million people in 1918, the number of deaths attributed to influenza and pneumonia rose and fell in the distinctive shape.[31] In the week preceding the second wave of the Spanish flu, 76 death certificates listed pneumonia or influenza (P&I) as the cause of death. By the end of the first week of this epidemic, that figure was nine times larger (706). By the end of the second week of the epidemic, the P&I mortality had climbed to 2,637 deaths; in the next week the total peaked at 4,597 deaths—a full 60 times larger than the week preceding the outbreak. As Farr had demonstrated in the influenza pandemic of 1847–8, in pandemics mortality rates climb in all categories, not just in the P&I figures.

Such a large number of deaths in such a short period of time meant it was difficult even to properly dispose of the flu's victims. Famously, as reported in an *American Journal of Public Health* editorial from October 1918, public health officials from the east coast of the United States (where the pandemic was raging) counseled their colleagues from the Midwest and West (where the pandemic had not yet appeared) that the first thing they should do to prepare for the imminent epidemic was to "gather up your wood-workers and cabinet makers and set them to making coffins." In addition, these health practitioners offered the sage advice that health department should put their "street laborers [to work] digging graves."[32] Sobering medical advice indeed.

An even more paralyzing task than disposing of the dead was caring for the sick. The sudden sharp rise in the patient load simply overwhelmed the medical capacity of communities of every size. Contributing to the difficulties in providing care was the fact that medical personnel were just as susceptible to infection as anyone else. Indeed, their constant interaction with the stricken made it more likely that they would be infected. In addition, many states had already stripped their medical services bare in order to provide physicians and nurses for their nation's efforts in the traumatic and destructive war that was still being waged into November. The crush of patients, a number of whom were desperately ill, swamped hospital facilities and clinics, with the incapacitated being lined up in the halls and on the floors between beds. Buildings were commandeered for temporary shelters and even public parks were appropriated for the erection of tent cities to house the ailing.[33] The sudden attack rate—perhaps combined with a healthy fear of the affliction—led to high absentee rates in a number of critical agencies such as public safety (police and firemen came in contact with many ill people as they transported them to the hospital); transportation (train conductors and others who had contact with large numbers of the public); and communication (telephone operators worked closely together in the exchanges). Normal life ground to a halt as the pandemic swept through.

Further amplifying the effects of the pandemic was the fact that it was a global phenomenon, reaching virtually every population on the planet at roughly the same time.[34] The simultaneous emergence of the second wave on three separate continents seemed diabolically arranged to maximize the efficient spread of the illness. From Brest the sickness rapidly afflicted the combatants on both sides of the trenches and soon radiated back to their respective civilian populations throughout Europe. From Boston the epidemic raced across the continent, speeding along the rail network of North America until it reached the Pacific. From there the illness was carried by steamers across the Pacific and down into Central and South America. From Sierra Leone Spanish flu journeyed inland, infecting sub-Saharan populations, as well as south by sea around Cape Horn and into the Indian Ocean interlinked system. As was the case for the Russian flu, the vast and speedy transportation web served as the charioteer for this highly infectious illness.

As the pandemic roared around the globe, medical researchers and scientists expended every effort to devise some protective or curative entity to deflect the terrible illness.[35] In the decades since the Russian flu pandemic, the germ theory of disease had conquered all its foes, with doubters either being silenced or shuffled off to the fringes of the medical debate. The prevailing wisdom was that Pfeiffer had uncovered influenza's cause, and therefore researchers around the world diligently searched for "Pfeiffer's bacillus," *Haemophilus influenzae*. Many of them found it either in colonies in the lungs of the sickened or in the oral/nasal tract. But some researchers had not found it. This was, at first, to be expected. "Pfeiffer's bacillus" was notoriously difficult to collect and to grow. But over time it became increasingly clear that the microbe was absent in a number of cases. This led some researchers to conclude that the sickness sweeping the continents was not influenza since "no influenza bacilli have been found."[36] Bacteriologists generally assumed that the testimony of those who failed to detect *Bacillus influenza* had made some sort of mistake in collecting, growing, or staining the elusive microbe.

Accordingly, as the pandemic gathered steam and accelerated around the planet, researchers rushed to isolate the organism and create serums, anti-toxins, and potentially even a protective vaccine to guard the public from the illness. Although not the hokums and magic elixirs peddled by the hucksters who appear at every unknown illness onslaught, these products from elite institutes and laboratories were probably no more effective. A torrent of Spanish flu neutralizers were sprayed, gargled, inhaled, swallowed, or injected, all to little effect. By 1919, it was increasingly clear that the inability of researchers to detect "Pfeiffer's bacillus" was not due to the technical short-comings of individual scientists, but a sign that not all people stricken by the flu hosted the bacilli. And if that was the case, then *Haemophilus influenzae* was not the agent responsible for the raging pandemic; nor was it the cause of influenza. Therefore, the multitude of pro-tective elements pressed onto the public by medical personnel were ineffectual in preventing infection by Spanish flu, and likely only of limited effectiveness in protecting against "Pfeiffer's bacillus," now seen as an opportunistic secondary invader. But if influenza was not caused by "Pfeiffer's bacillus," what was its cause? That remained an open question.

The second, killer wave, of Spanish flu encircled the globe in the fall and early winter of 1918–19. This second wave was followed by a third wave, which rebounded around the planet again. Although much less deadly than the second wave, some places were hard hit by this deadly return. The third wave appeared milder only in com-parison to the disastrous second wave. Indeed, if it were not for this catastrophic second wave, the third wave of Spanish flu would have been an unprecedented pandemic disaster. A further, milder, fourth wave of the pandemic appeared in a few places in early 1920 before the Spanish flu pandemic was spent.

In the wake of the pandemic, medical researchers and scientists assessed their responses to combating it and were forced to conclude that their efforts had been useless. Research personnel realized that nothing they had tried had effectively deterred or deflected the pandemic, and for that matter, had not conclusively protected even one individual. The medical establishment determined that only palliative nursing care had offered any benefit and even this basic service had been in short supply at the height of the pandemic. Efforts at studying the illness had not protected the public; indeed, the unaltered course of the pandemic had clearly demonstrated that the research community did not know as much about the illness as it thought it did. Prior to Spanish flu, researchers could at least say they knew what caused influenza. In the wake of the calamity, that starting point of research had been swept aside. The medical community may not have known what caused influenza, but they were generally in agreement that it was not "Pfeiffer's bacillus." Some researchers suggested that the illness was caused by a "filterable virus"—*virus* meaning "poison" in Latin—that was invisible to microscopes and impervious to staining. But this idea remained mere hypothesis and filtering experiments had been inconclusive.

The aftermath of the disastrous epidemic represented a crossroads in studying influenza. Since the medical establishment's fundamental understanding of the illness was faulty, there was then a need to conduct research to understand the disease at the most basic level, beginning with identifying the causative agent. The stunning number of deaths and illnesses demonstrated that influenza was no mean threat. Therefore, there was a pressing demand for research. Some researchers pointed out this fact and issued a call for renewed investment and interest in studying the mysterious affliction. For example, U.S. Surgeon General Rupert Blue envisioned a government-funded institute dedicated to unraveling the disease's secrets.[37]

Blue was to be disappointed. No great institutions were created to investigate influenza, and at the institutes that already engaged in medical-related research (for example, the Rockefeller Institute), influenza was not a major topic of inquiry.[38] Instead, study of the disease became largely an individual pursuit, and, to some degree, almost a lonely one. There was some public and private research support—for example, the British Medical Research council sought to do some basic research on influenza's cause—but again, this effort was piecemeal.[39]

Instead of serving as a spur to renewed and vigorous research into the disease of influenza, medical researchers and the public adopted a policy of institutional "forgetting." This amnesia did not extend to the individual level, as the searing experiences of the sickness and death of loved ones remained indelible, but, as Alfred Crosby pointed out, the terrible event was not commemorated in public memory. There were no statues dedicated to the sick; the pandemic did not figure prominently in the biographies of major figures from the period; and the Spanish flu was generally not included in the histories written about that time and the war.[40] This "forgetting" was especially prominent among the medical researchers and physicians who had been charged with taking care of public health during the war. Doctors in the U.S. Army studiously overlooked the terrible mortality toll of Spanish flu in recounting their efforts to protect the military against infectious diseases during the Great War.[41] Although one cannot speak conclusively about the myriad motivations of individuals, it seems that the distinct failure of the medical establishment in the case of the Spanish flu did not fit the triumphant narrative of medical research in the early twentieth century, and so it was therefore ignored.[42]

Whatever the rationale for the long-term omission of the pandemic event from the public consciousness, and for the relative dearth of institutional support for influenza research in the interwar years, there was one topic that researchers sought to investigate in the immediate aftermath of the pandemic: How many victims were there? Typically, as we have seen, influenza was deadly only for the very young or the very old, or for those whose lives were already under assault from some other affliction or condition. Even those who succumbed to influenza were few and far between. Investigating Spanish flu's toll would reveal that this pandemic had upended the usual mortality pattern.

Spanish Flu Accounting

When the second wave of Spanish flu arrived in a community, it generated so many patients—and so many of them so very ill—that the medical infrastructure of a community was rapidly overwhelmed.[43] Medical personnel, if not stricken themselves, were burdened with dozens and then hundreds of sick patients needing care. The immediacy of the health emergency meant that the bulk of physicians and even researchers were pulled into patient treatment, which generally precluded large-scale studies or larger perceptions of the flu's impact. Instead, physicians and researchers wrote up their experiences in their clinics and hospitals, and their impressions from what they saw there. Such reporting provides a snapshot of the pandemic. One of the impressions frequently noted by these doctors was that the Spanish flu seemed most deadly and devastating to the youngest and previously most robust members of a community. Doctors noted that fit, young men and women were overrepresented in their autopsy rooms and morgues. But, as these reports often noted, the authors did not have many opportunities to compare their observations with those of others.

Snapshots are useful in that they freeze a moment in time for later study and reflection. But the freeze-frame is limited in that it only depicts one moment in time at one location. Ideally, researchers would like to have a variety of snapshots from a diverse set of locations, for one alone is an anecdote. Anecdotes are useful and suggestive, but not conclusive. A collection of multiple scientific observations is data, and data can be used to provide a more comprehensive mosaic from a variety of points. As the second and third wave receded, public health officials and researchers began to assess the impact of the influenza pandemic in the fashion Farr had taught them: to begin to count up the number of extra dead. The evidence collected supported the observations made by a number of sources: the younger members of the community had suffered an unusually high morbidity and mortality impact; unusual, at least, for influenza. Indeed, those who generally suffer the lightest touch from an influenza epidemic— those in the 20–40-year age bracket—actually suffered the highest rate of mortality in the Spanish flu. This anomalous point is readily captured in a simple age vs mortality graph.

Plotting the victims of a typical influenza season on an age vs mortality graph generates a common pattern. Those with the weakest immune systems—the very young and the very old—suffer the vast majority of deaths. On such an age/mortality graph you have a clustering of deaths in the 0–5-year age range, which drops steeply to a very low plateau that stays generally flat. The line of the graph begins to rise slowly through the age brackets of the thirties and forties, and then rises sharply from the age of 60 onwards. The resulting graph looks like a large "U." Such a mortality pattern can be gleaned from as far back as our records allow, as ancient chroniclers of events that historians have later ascribed to influenza epidemics often noted that only some old folks or the very young were at mortal risk. For that matter, the pattern is largely the same in the present day. Spanish flu, however, did not follow this mortality script. Instead, the largest numbers of deaths were recorded in the 20–29-year age category, with the 10–19 and the 30–39 brackets not far behind. Graphing the deaths of the Spanish flu along that age/mortality axis returns a pattern of a crude "W." Mortality in the youngest age grouping was high, but this peak was surpassed by the clustering of victims in the 10–40-year age range. Curiously, the aged columns (age 60 and above) show a disproportionately lower percentage of mortality from the Spanish flu, an oddity that presumably also extended to morbidity as well. The picture of Spanish flu's victims was the same everywhere around the globe, as reported by physicians and states in their observations and data.[44] No other influenza pandemic before or since had such an unusual pattern.

Answering the question of who the Spanish flu killed leads directly into the question of how many. This is much more difficult to answer, for the pandemic was a fast-moving global affair. The steep rise of cases rapidly swamped even the most prepared medical facility. At the height of the pandemic, one of the first things to be left by the wayside was the paperwork aspect of hospital administration, either because of the dimensions of the emergency or because of the shortage of healthy (and willing) personnel. In a number of locations, the records fell silent as the crisis reached its apex. In addition, bookkeeping would only account for those that were admitted into formal medical settings. An uncounted—and uncountable—number of people were forced to face the illness at home, whether by proclivity or the fact that hospitals were completely overwhelmed. These difficulties in tracking patients would make the accuracy of any accounting suspect, but these problems were comparatively minor in assessing Spanish flu's total impact.

Spanish flu was a global pandemic and if those states with a long tradition of bureaucratic infrastructure and strong public health systems were inundated, those areas without this administrative background presented even greater difficulties for the mortality tallier. Despite the number of obstacles that measuring the mortality of the pandemic presented, a number of researchers have accepted this demographic challenge over the years.

The first serious large-scale effort to calculate the death total was undertaken by Edwin Oakes Jordan at the behest of the American Medical Association.[45] Jordan assembled a number of medical reports and bulletins from around the world to try and assess the course of the pandemic. From these reports he extracted a general overview of the pandemic's devastating jaunt around the globe. Jordan also collected as many government and public health censuses as he could in order to generate case fatality rates, attack rates, and, ultimately, a global total of mortality. In some places, the data was relatively intact, and he needed only to extrapolate rates from one location to another; in others, the data was patchy; in still others, it was nonexistent. From this mixed bag, Jordan estimated that 21.6 million people died from the combined effects of the various waves of Spanish flu.

Whether out of respect for or intimidation by Jordan's research and his systematic efforts, the global estimate of 21.6 million was accepted for decades, despite the fact that more in-depth demographic studies had begun to challenge his estimates. For example, the great historical demographer Kingsley Davis had found a curious void in his population estimate for the Indian subcontinent over the decade of 1911–20. Writing in 1951, Davis discovered that the census figures for 1920 reported a number significantly smaller than would have been expected based upon the demographic trends of the previous censuses. After determining that the census figures were reliable, Davis applied the average of the population growth from the preceding and following decennial censuses to the census of 1910. Davis discovered that there were approximately 20.5 million people missing from the census of 1920. The only major public health event in that decade was the Spanish flu, prompting Davis to conclude that it had killed somewhere around 19 million people. He judged that the additional 1.5 million shortfall in the census could be attributed to the number of children not born to those who had died in the pandemic.[46] Davis's total just for India and Pakistan almost equaled Jordan's global total. Despite that fact, and the fact that other studies based upon numbers and models unavailable to Jordan were being produced for a variety of regions, Jordan's global estimate remained mostly unchallenged into the 1990s.

In the 1980s, a number of more precise demographic investigations into the Spanish flu outside of Europe and North America began to appear. These localized studies generated a sturdier set of figures related to Spanish flu's mortality impact, and served to fill in gaps of information where Jordan had been forced to speculate. These richer, more accurate studies forced a long-overdue reassessment of Spanish flu's mortal toll. The task was taken up by prominent medical historian K. David Patterson and his frequent collaborator, medical geographer Gerald Pyle.[47] Drawing upon their own work investigating the demographic impacts of the pandemic—notably on the continent of Africa—Patterson and Pyle created a work of "global synthesis" that drew upon the numerous demographic studies of regional population patterns. Based upon the evidence available to them at that time, Patterson and Pyle offered that mortality probably ranged between 24.7 million and 39.3 million deaths. They placed their most likely estimate at approximately 30 million. A decade later, Niall P.A.S. Johnson and Juergen Mueller revisited the estimation process again in an article of 2002.[48] Relying upon new studies and models of influenza pandemics, Johnson and Mueller estimated that the true total of the Spanish flu was approximately 50 million. Remarkably, Johnson and Mueller claimed that their conservative estimate could be undercounting by as much as 100 per cent, meaning that it was possible that Spanish flu claimed 100 million victims around the world.

Aftermath

The Spanish flu was devastating. It was demoralizing for bacteriologists and germ theory proponents, who were forced to abandon *Bacillus influenza* as the organism responsible for influenza; demoralizing for physicians and research scientists, who realized they had had no effective treatments for their stricken patients; demoralizing for public health officials, who had been unable to protect their charges as the pandemic arrived, crested, and receded unaltered by their ministrations; and most demoralizing for the people who watched many of their loved ones dying feverish, choking deaths. Apparently, it was so demoralizing that the medical and scientific community threw up their hands in despair of ever solving the riddle of the disease.[49]

As the last pandemic wave of Spanish flu receded, and the influenza disease returned to its usual seasonal pattern of being a minor irritant with a low death toll, research interest in it dwindled. Whether that was because of frustration at the difficulties in even developing a basic understanding of the illness or because of the enticements of research in other, more lavishly funded research programs is impossible to know. What we can say is that influenza research did not receive much institutional support in terms of either funding or resources. Research on influenza was often an individual pursuit, undertaken for the most part in isolation and with funds earmarked for other projects. These individual researchers began to unravel the mysteries of the disease, beginning with identifying its causative agent. From these breakthroughs, influenza research finally began to attract the support necessary to expand the knowledge of the disease. International collaboration led to further insights and eventually to the creation of protective methods. Jordan was correct in his prediction—pandemic influenza did appear again in the twentieth century—but public health organizations were armed with new knowledge about the virus and how to prevent it. But lack of knowledge—and the ability to effectively and speedily act on it—would continue to hobble tactics designed to protect the public from pandemic influenza. These failures, however, did not arise from lack of effort. As we shall see, in the twentieth century, two pandemics, one anticipated pandemic, and one accidental epidemic prompted large-scale programs to interdict or dampen the deadly spread of influenza. Although in subsequent years influenza would not kill anywhere near the total claimed by Spanish flu, the illness was certain to assault the public, causing sickness and death every year. As scientists developed techniques to control and even eradicate a number of killer infectious disease during the twentieth century, influenza stubbornly remained outside of the control of medical approaches. The difficulties influenza posed prompted some researchers to call it the "last plague." This last pandemic menace remained a vexing challenge into the twenty-first century.[50]

FOUR

The 1920s to the 1980s

Influenza research in the decades following the Spanish flu was marked by a vision problem; initially, quite literally, as not knowing either which causative agent to look for or how to look for it hampered influenza research. After the element responsible for the illness was correctly identified as a virus, a question arose: What could be done about the disease? The most pressing need was to prevent an epidemic, or, failing that, to protect the public. Vaccination was seen as a promising avenue in safeguarding people from the illness, and vaccine research—which in the post-Second World War period began to draw the funding and support of national governments—made some tremendous breakthroughs. But the mutable nature of the virus soon rendered these doses ineffective and influenza researchers realized that they would now have to be able to see very quickly if the virus was evolving away from the protection of the shots. New understanding of the virus brought new efforts to control it, such as the development of a global surveillance system. The goal of this global surveillance system was to capture new strains as they appeared and began to circulate, with the intention of being able to match vaccines to these new viral types and so prevent illness.

Protecting against influenza presented a number of difficult challenges to scientific researchers. The phenomenon of viral drift meant that public health professionals had to be vigilant in watching for the appearance of newly mutated strains that made the vaccines in use less effective. Over time, increasingly sophisticated measures of sickness and death revealed to these medical officers the true costs of influenza to a population. But, even though public health workers appreciated that influenza had high costs to a society in terms of both lives and money, it was not the seasonal appearance of the illness that was their greatest concern. Rather, it was the sudden appearance of a pandemic that health experts sought to guard against. The Russian flu epidemic of 1889–92 was very disruptive of communities as it swept through in pandemic waves. Commerce, government, and everyday life had halted as a significant number of citizens were stricken by the illness. A number of those assaulted by the affliction died. While only forming a small percentage of the infected, the widespread nature of the illness generated a sizable death toll; and the impact of Russian flu was dwarfed by the catastrophe that was the Spanish flu. It was for these influenza pandemics that medical officers remained alert.

Influenza, however, remained a confounding problem for scientific researchers. While continual work on the virus unraveled many of the mysteries of its behavior, and the expanding network of surveillance provided information that meant scientists could continue to calibrate vaccine protection, there had not been another pandemic in the decades following Spanish flu. No researcher familiar with the history of influenza could have any confidence that influenza epidemics were a thing of the past; indeed, the more one searched the historical record, the more one came to the conclusion that a new influenza epidemic was certain to appear. Public health officials had developed a plan to protect against an influenza epidemic: a global system of surveillance tied to the national production of vaccines. But this plan had not yet had a chance to be put to the test. This pandemic-free period was soon to end, however, as two pandemics would emerge over the course of a little more than a decade in the 1950s and '60s. In both cases this system of surveillance and vaccination failed. Public health officials had been able to "see" the appearance of a pandemic strain, but the new virus had moved too quickly for them to effectively develop, produce, and distribute a vaccine. Successful interdiction of a pandemic would require rapid action upon the appearance of a new viral type. Effective rapid action would require prior planning and preparation, but such a hair-trigger alert carries its own dangers.

The speed of the influenza virus's transmission required accelerated decision-making based upon the first appearance of the new strain. The need for quick assessments and for the hasty development of effective vaccines led to two unforeseen events in the late 1970s. In the first, a massive vaccination program was launched to protect against a pandemic that did not appear. In the second, study of the virus started an accidental man-made epidemic. The combination of failed responses and unintended viral releases prompted a retreat from active pandemic response preparation. Concern over influenza pandemics would recede again from the agendas of public health planners, before returning to the fore in the closing decade of the twentieth century.

The Cause of Influenza—A Vexing Question

The failure of medicine in the face of Spanish flu had thrown the identification of the causative agent into disarray. Many researchers concluded that "Pfeiffer's bacillus" was not the element responsible for the affliction and was, at best, a prominent member of the cohort of secondary invaders. But the connection between the bacillus and influenza died hard. One could at least *see* this microbe. Indeed, a number of medical texts were still reporting that influenza was caused by *Haemophilus influenzae* into the 1930s. But among scientific researchers, this idea was falling increasingly out of fashion in the 1920s.[1]

Medical scientists had abandoned the Pfeiffer explanation for influenza, but this left open the question of what the cause of the disease was. The initial solution seemed to be some filter-passing poison, or "virus." Viruses were mysterious entities. Research on the tobacco mosaic virus had conclusively proven that there was some element smaller than a bacterium that could cause an illness that could be transmitted between living organisms.[2] Indeed, some British researchers had been positing that a virus was the source of influenza since 1914, but being able to demonstrate this was proving very difficult.[3] The method of demonstrating a virus to be the cause of an affliction entailed screening the material through very fine porcelain filters to eliminate any potential bacterial sources of infection. Probably because of the crudity of the filters, a number of experiments testing the viral concept of influenza onset produced a jumble of results; some supported and some eliminated a virus as the source of influenza.[4] The upshot was that critics of the viral theory of influenza charged their opposites with using virus as an "alibi for their failure" in demonstrating the entity responsible for influenza.[5]

The difficulties in conclusively identifying the element responsible for influenza frustrated medical experts. During and after the Spanish flu catastrophe, a number of researchers around the world sought to pin down the agent responsible for the pandemic. The riddle of influenza required focused research to answer the basic questions of the epidemic threat, but as the 1920s continued, fewer scientists were grappling with these questions, and those researchers who were working on them had little institutional support. However, it was to be those individuals who picked up the research thread of a viral source for influenza who would make the first dramatic research breakthroughs on the virus.

Proving the Viral Origin of Influenza

Richard Shope was one of those scientists who took up the issue of uncovering the source of influenza in the late 1920s. Shope, a native of Iowa, was working at the Princeton Laboratory of the Rockefeller Institute when he decided to attack the problem of the identification of the virus in an indirect way. He was aware of a curious report from the Spanish flu pandemic made by an Iowa veterinarian named J. S. Koen, who had been working for the United States Bureau of Animal Husbandry during the time of the pandemic. Koen had observed the outbreak of some infectious disease he believed was new to the pig population. The epidemic produced symptoms of sneezing, nasal discharge, fever, and prostration. Moreover, the outbreak in the pigs trailed the arrival of the Spanish flu pandemic in the human population very closely. The similarity of symptoms in the pig and human populations and the close proximity of the swine sickness to the arrival of the human pandemic prompted Koen to assert that the pig and human diseases were one and the same. What was especially tantalizing about this relationship between pig and human influenza for Shope was that whereas the Spanish flu had apparently disappeared from humans, the illness, new to the pigs in 1918, reappeared in swine herds every autumn. It would emerge with the onset of cooler weather and quickly move through the swine populations of the American Midwest, prompting the same symptoms as had been seen in previous years.[6]

Shope decided to travel to the Midwest and collect samples from sick pigs with the intention of passing the filtered material to a healthy pig in an attempt to induce the illness. In so doing, he would demonstrate that the sickness in pigs was caused by a virus; a pig ailment, he believed, that was generated from human cases of Spanish flu. After gathering samples from infected herds in the Midwest, Shope returned to his laboratory, where he passed the gathered material through state-of-the-art bacterial filters. Then, using a technique recently demonstrated to pass infections between pigs, Shope dripped this filtered material into the snouts of the animals. His experiments were a success. Shope was able to serially transmit the same disease after filtration from sick pigs to healthy ones, proving it was caused by a virus. In 1931, he published his breakthrough experiments in the *Journal of Experimental Medicine*.[7] But, as Shope's critics pointed out, Shope had only successfully demonstrated that the pig disease was caused by a virus; he had not yet proven that the human disease of influenza was caused by the same, or any, type of virus.

Meanwhile, across the Atlantic, a cadre of British researchers had continued to pursue the supposition that influenza was caused by a viral contagion. These researchers worked at the National Institute of Medical Research (NIMR), a division of the Medical Research Council (MRC). The MRC, founded prior to the outbreak of the First World War, had grown tremendously as a result of the militarization of medicine during the war. In the aftermath of the conflict, the MRC sought to carve out a research niche of its own. One of the avenues it aimed to investigate was the diseases caused by filter-passing agents (viruses). It tapped the NIMR as the agency responsible for these research efforts.[8] Under the direction of Patrick Laidlaw (later Sir Patrick), the British government provided modest sums for equipment and personnel for research on viruses, including the possible viral origin of influenza—demonstrating that the Spanish flu pandemic was not completely forgotten. A closer look at the NIMR's research work on influenza, however, reveals a rather parsimonious level of support.

Laidlaw attracted a small group of young researchers to investigate filter-passing diseases, including Wilson Smith and C. H. Andrewes. While the MRC had been able to maintain a steady stream of research dollars during the 1920s, no small feat in the straitened economic climate of Britain at that time, the funds were spread across a variety of virus-related investigatory programs. In fact, the NIMR was not directly researching influenza at all by the early 1930s. Instead, it was conducting proxy research on distemper in dogs. There was a sizable quantity of research funding available for research on distemper, primarily supplied by dog enthusiasts, who donated money for research on the puzzling canine affliction. These monies were solicited through *The Field*, a magazine for the country gentleman. These private funds had initially supported dog-breeding colonies for the development of experimental vaccines, but the dog populations were soon replaced by ferrets, which, it had been discovered, were also susceptible to distemper. The use of ferrets turned out to be a serendipitous choice for influenza research.[9]

In 1933, a steep epidemic of influenza broke out in England. Smith and Andrewes seized on this opportunity to determine whether the illness was caused by a virus. The first step was to introduce the infection into an experimental animal for research purposes.[10] Smith and Andrewes took filtered samples from influenza patients and injected the material into a host of laboratory animals (curiously, not including ferrets), all without success. In the wake of these failed experiments, they received word from their colleagues at the Wellcome Institute that the ferrets they used for distemper research had apparently contracted an illness that mimicked the symptoms of influenza. Inspired by this news, Smith decided to try to infect ferrets with influenza. Smith took a throat-washing sample from Andrewes (who had conveniently come down with a case of influenza), filtered the material, and, drawing upon Shope's example, opted to drip the material into the noses of the ferrets. Within days the ferrets were sneezing, feverish, and lethargic, apparently manifesting an influenza infection. Smith and Andrewes were able to pass the material from sick ferrets through bacterial filters and infect healthy ferrets, demonstrating the infectious element was a virus. When the chain of ferret transmission was accidentally broken, Andrewes was able to re-establish it with washings from Smith, who now was suffering his own bout of the flu. This sample was eventually stored as the W. S. strain, which became a laboratory staple. Subsequently, the infection circulating in ferrets was transferred back to a human when a sick ferret sneezed into the face of a laboratory assistant (Charles Stuart-Harris, who went on to a prominent research career and was also honored with a knighthood).[11] In an ironic footnote, it turned out that the initial ferrets at the Wellcome Institute that had sparked the suggestion that ferrets might be susceptible to influenza did not have flu at all; rather, the animals had contracted distemper.

Smith and Andrewes's work with the ferrets was a tremendous breakthrough. Not only had they conclusively demonstrated that influenza was caused by a virus, they had also discovered a laboratory animal suitable for conducting research on the virus, accidentally or not. The ferret is an excellent stand-in for humans in studying influenza, since the virus behaves in similar fashion when it infects the respiratory systems of both ferrets and humans. Within a year Smith and Andrewes had developed a technique to infect mice with the virus, which opened up further research possibilities, since mice are cheap and plentiful. Wilson Smith contributed a final piece to easing influenza research by demonstrating that the newly discovered virus replicated very well in developing chicken eggs.[12] The discoveries at NIMR opened the floodgates for influenza research. The illness was proven to be caused by a virus, a detailed animal model (ferrets) for human influenza had been discovered, a cheaper and more plentiful animal (mice) had been identified for use in studies, and an inexpensive and efficient viral production system was developed (fertilized chicken eggs). Quite a few feathers to put in any laboratory's cap.

The new information about influenza galvanized researchers around the globe, especially in the United States. Some of these scientists had already begun work based upon the supposition that the illness was generated by a virus; after all, that was what Shope believed he had demonstrated. But the results produced by Andrewes and Smith charged this research area. Scientists on both sides of the Atlantic Ocean transferred information to one another through both published papers and private correspondence.[13] These researchers also made a habit of periodically visiting one another's laboratories to demonstrate techniques and to swap ideas. Research work on influenza, which had stagnated, suddenly took off.

After discovering the viral origin of the infection and developing methods to study and then produce the virus, the next logical step was to develop some protective or curative technique. The obvious method was to produce a vaccine. But creating a vaccine required solving some basic questions. What should be put in the vaccine? How much of it? How could one determine if the vaccine had worked and the individual was protected? Answering these questions would take time, money, and resources. As it would turn out, British researchers would find their access to all three of these constricted with the onset of the Second World War. Conversely, researchers in the United States would discover that their opportunities to receive resources and money (at least) would be greatly expanded because of the war.

Developing a Vaccine

The first large-scale attempt to create a protective vaccine against influenza was undertaken at the same place where the virus was discovered: the laboratories at the NIMR.[14] Smith, Andrewes, and their collaborators sought to create a protective vaccine using viral material generated from mice lungs and inactivated (or killed) using a formaldehyde product. The resulting material was filtered to remove impurities and injected in small-scale tests. Although the initial tests produced satisfactory antigen reactions—suggesting that the vaccine was protective—subsequent larger field tests failed. The reasons for the vaccine trial failures are multifaceted and complex, but a few key elements can be readily described. First, even after filtration, using the material from the mice generated strong reaction rates in the inoculated. The injection of foreign materials into the bodies of people prompts the immune system to react to the presence of proteins and other material mixed in with the vaccine to be inoculated, resulting in fevers, swelling, abscesses and even more dire shock effects. Harvesting viral material from mouse lungs contained extraneous material from the animals that proved difficult to remove prior to injection.

Second, there was no efficient way to measure how much of the viral material was in each dose.[15] Finally, there was also some bad luck involved in detecting the effectiveness of the vaccine. Since the only way to measure whether the vaccine was effective was to see if the inoculated were protected against the disease, the researchers tried to time their trials to start right before the onset of influenza season. In early 1937, experimental injections were just getting under way when an influenza epidemic appeared. The vaccine did not offer any protection, but it was possibile that the people had been infected before the vaccine could produce antibodies.

Researchers in the United States continued to work on safeguarding inoculations through the late 1930s. And, in a similar fashion to their British compatriots, the American researchers found the results of the studies to be inconclusive, with unacceptably high adverse reaction rates to the vaccine and unclear protective effectiveness. A safer production model would need to be developed, as well as a quicker, more efficient antibody detection system, if a vaccine was to be used on a large scale.

Solutions to both these problems would appear in the next few years. In 1940, the Australian scientist Frank Macfarlane Burnet (later Sir Frank) perfected a method of producing large quantities of viral material in fertilized chicken eggs. Researchers had relied upon mice (and some ferret) sources to produce vaccine material because they proved more efficient for large-scale production compared to the egg method developed by Wilson Smith. Burnet's new technique fostered massive production of viral material using chicken eggs that was cheaper, faster, easier, and more efficient than using mice. Egg-produced vaccines also had significantly lower reaction rates in the injected than animal-produced vaccines. In addition, in 1941, the American scientist George Hirst developed a rapid test to determine the antibody production rate of serum samples that was both simple to use (one could readily see the reaction in a test tube) and could be conducted quickly. Both scientific discoveries held implications for large-scale vaccine production and use.[16] These manufacturing breakthroughs came too late for the British scientists to utilize them, however. Wartime rationing made acquiring eggs impossible for researchers and the conflict had generally restricted resources for the laboratories.

The American researchers, on the other hand, did not face the restrictions that British laboratory workers faced because of the war; in fact, just the opposite occurred. The conflict led to massive state investment in the sciences, which opened up opportunities and resources for scientists and medical researchers. Protecting against influenza became one of the primary concerns of the u.s. military, which had taken steps to protect troops against pandemic influenza even before the official u.s. entry into the war. In January 1941, the u.s. Army created the Board for the Investigation and Control of Influenza and other Epidemic Diseases in the Army.[17] The Board was a remarkable collection of scientific talent, and many pioneers of virological research served under its auspices, including Jonas Salk, Albert Sabin, and John Paul. The Board was under the direction of Thomas Francis, who was at that time the premier American researcher on influenza.

Francis had begun his influenza research work at the Rockefeller International Health Division. From there he left to join New York University and eventually wound up at the University of Michigan. While at NYU, Francis announced that he had identified a completely different type of influenza, which he dubbed Influenza B. Later, two research groups simultaneously identified a third, much rarer, type of influenza, called Influenza C.[18] The charge from the Board for Francis and his team was to develop a protective vaccine against influenza for use among troops during the war. Francis tackled the problems that had hindered previous influenza vaccines, which he identified as a variable level of viral material per dose—it was difficult to determine how much inactivated virus was in each dose; high reaction rates among the vaccinated—marked by high fevers, sore arms, and occasionally worse shock effects; and slow production methods—because they relied upon laborious harvesting of mouse lungs. The key solution to many of these problems was to concentrate and purify the viral material. Initially, a variety of techniques were used to achieve this consistent vaccine material, including a process of freezing and thawing, but eventually the most effective means proved to be using high-speed centrifuges, which spun out miscellaneous material, leaving only the concentrated virus behind. Although later experiments suggested that there was a boundary level beyond which too much viral material produced high reaction rates, during the war these concentrated vaccines generated very high protective antibody rates, demonstrating that a vaccine could safely protect against a circulating influenza strain.[19]

The interest in and support of influenza research by the military provided Francis with greatly expanded resources to conduct his experiments. It also provided him with access to an excellent population for controlled vaccine studies; a population that was young, healthy, and readily tracked. In 1942, Francis and his team undertook a fair-sized trial among volunteers at Cornell University and two institutions in Michigan. The group inoculated one population while leaving the other as a control. Unfortunately for the experimenters, 1942 proved to be a very mild influenza year, so there was an insufficient number of cases to determine whether the vaccine had been protective or not. The scientists planned an even larger test for the following year. Nearly 12,500 people were enrolled in a study involving a number of university campuses from New York to California. The majority of these "volunteers" were soldiers and sailors attending the universities for service in the military; they were thus readily traced and compliant with the experiment's demands. Half of the group was inoculated and half left as unvaccinated controls. This year the virus cooperated: 1943 was marked by a sharp epidemic around the United States. As the researchers tallied the results, it became clear that the vaccine had worked. The unvaccinated were between three and four times more likely to contract influenza; on some campuses, the ratio of unvaccinated influenza sufferers to vaccinated ones was as high as six to one. In addition, there was a low rate of adverse reactions in the vaccinated group. By the end of the war all military personnel in the U.S. and many Allied forces were receiving protective influenza shots. Shortly thereafter, influenza vaccines were licensed for general use. Protection against influenza pandemics seemed to be close at hand.[20]

The Vaccine Failure of 1947 and the Creation of the World Influenza Centre

The vaccine injected into the military volunteers in the winter of 1943 was actually a trivalent vaccine containing viral components from two type A strains (PR8 and Weiss) and one type B strain (Lee).[21] In succeeding years the recipe remained the same, and the vaccine maintained its high degree of protection until 1947, when it failed completely. This vaccine failure would shake the confidence of public health officials that they had an effective tool for protecting the public from an influenza pandemic. But from this failure would come increased knowledge of how the virus works; it would spur the development of a global surveillance system designed to track emerging pandemics.

Early in 1947, a sharp epidemic of influenza appeared and circulated rapidly across North America. Francis had continued his vaccine research, and had undertaken another large study through the University of Michigan Student Health Services. The study had inoculated a subset of the population of students and had created a tracking program for students who had declined the influenza vaccine. As the epidemic enveloped the campus, Francis and his collaborators totaled the number of likely influenza cases from the pool of the vaccinated and unvaccinated alike. To their surprise and disappointment, the rate of influenza infection was virtually the same in both groups (7.19 per cent in the inoculated contracted influenza as compared to 8.09 per cent in the unvaccinated). The vaccine had failed.[22] The question was: Why? Clearly, the vaccine did not match the circulating influenza strain that had precipitated the epidemic. The American researchers had been hopeful that the number of influenza strains were limited and would therefore be vulnerable to a few protective shots. Unfortunately, it was now becoming clear that influenza strains could change. The scientists had just discovered the phenomenon of viral drift and viral shift.[23]

Whereas American researchers were surprised by the changing virus, the appearance and rapid circulation of the new strain in 1947 served to confirm some ideas Christopher Andrewes had been developing about the nature of flu viruses. Andrewes had been closely tracking an Influenza B epidemic that had begun in June 1945. While not as explosive as an Influenza A epidemic, B strains can spread widely and cause a crisp uptick in the number of infected. The new B strain appeared to have emerged in one location (the Pacific) and then spread around the globe. In similar fashion, the new Influenza

A strain (subsequently called FM1) was first seen in Australia in 1946; by 1947, it had circulated around the entire world. Combined, these two examples suggested that the influenza virus is an unstable entity that can change. Those changes appear in one place, then proceed to transit around the planet. No one vaccine will be able to provide protection, as the new strain will render the protective shot ineffectual. In order to be operative, an influenza vaccine would need to be matched to the prevailing type of virus, meaning that the vaccine must be patterned on the new strain. Detecting these new strains required vigilance, and since the new strain could appear anywhere, such vigilance would have to be global. Andrewes concluded that the world needed a global influenza surveillance system.[24]

The natural place for such a global surveillance program was the newly formed World Health Organization (WHO), the health arm of the United Nations. Andrewes approached the organizing committee of the United Nations and suggested that global influenza surveillance be one of the tasks of the new health program. They readily agreed. In September 1947, even before the WHO was officially ratified, the World Influenza Centre was open for business. Housed in borrowed space at the National Institute for Medical Research in London, Andrewes set up a typing laboratory for influenza strains. The design of the program was simple: scientists around the world would send in samples of suspected influenza strains circulating in their communities that they had been unable to type. Andrewes's laboratory would quickly determine the make-up of the samples and, if circumstances warranted it, create a vaccine-ready seed stock for the new strains. The structure of the system was modified almost immediately as American scientists chafed at the subordinate role assigned them by Andrewes. American money and expertise were crucial for the success of this surveillance system, so a co-equal typing laboratory was set up in the U.S. and given responsibility for identifying novel strains in samples collected in the Americas.[25]

From the outset the system was designed to be more than just a study program. It was meant to be an active part of protective public health. Once a strain was determined to be novel and one that the current vaccine did not protect against, the reference laboratories in the UK and U.S. would create vaccine seed stock that would be distributed to manufacturers. The plan was to quickly produce and distribute a vaccine to protect the public. The effectiveness of such a public health structure relied on rapidly detecting new influenza types, which in turn relied on the number and reach of laboratories gathering samples around the world. Although initially quite modest and largely centered on Europe and North America, by 1953 there were two reference laboratories (London and Bethesda, Maryland) and 54 designated influenza centers in 42 nations.[26]

While the surveillance system was global in intent—utilizing national resources (laboratories and personnel) to achieve WHO goals —vaccine manufacturing and distribution would be strictly national affairs. The WHO had no funding or capacity to create vaccine manufacturing sites and it considered its role to end once the vaccine seed stock was generated for new influenza strains. This dichotomy between the global approach to surveillance, which to be effective needed a wide net of laboratories in a multitude of countries, and national vaccine production and distribution, limited to a handful of states with the means to produce, purchase, and distribute protective inoculations, remains a problem to the present day.

By the mid-1950s, the manufacture of influenza vaccines relied on the use of fertilized chicken eggs in methods that developed the technique pioneered by Burnet. The process of production required that the seed virus be introduced into the egg and incubated for a few days. The egg was then cut open and the allantoic fluid removed.[27] Generally, only one or two vaccine doses could be produced by each egg. The difficulty in manufacturing vaccines from chicken eggs was not so much the technology of replicating the virus, although sterility needed to be maintained throughout the entire process (a potentially tricky task when a living biological entity like a fertilized chicken egg is harvested by human hands), but in the scaling-up process for mass production. Manufacturers would need to have access to very large numbers of fertilized eggs, incubation facilities, and centrifuges in short order to make any sizable quantity of protective doses.

In order to make enough of the vaccine for a large population, manufacturers would need time. That was precisely what the surveillance system was designed to give them. Andrewes's program was created to give an early warning so a vaccine designed to protect the public could be produced in time to blunt an emerging pandemic. But since the system's founding in 1947, there had not been a pandemic to which it could respond. That was coming; and it was a test that the system would fail.

The Asian Flu of 1957

In mid-April 1957, Maurice Hilleman sat reading the New York Times in his office outside Washington, DC.[28] The office was not an ordinary business office; nor was its reader an ordinary New York Times reader. Hilleman was a scientist at the Walter Reed Army Institute of Research (WRAIR) and he was tasked with the study of respiratory illnesses of military significance. At the head of that list was influenza. While the United States had enthusiastically joined in with the WHO Influenza Surveillance System, it had also maintained its own military-based surveillance system. Hilleman sat at the center of a sentinel system linked to military deployment of U.S. forces around the world. With the global dispersal of U.S. forces in the wake of the Second World War and the onset of the Cold War, that meant that suspect samples of potential epidemic diseases were shuttled to Hilleman's laboratory for identification and study from a number of regions. It meant that Hilleman had a global reach potentially equal to that of the WHO system.

The article that caught Hilleman's eye was a small four-paragraph "filler" in the 17 April 1957 New York Times edition headlined "Hong Kong Battling Influenza Epidemic." The brief report stated that approximately 250,000 of the colony's 2.5 million citizens were receiving treatment for an illness identified as influenza. Hilleman later reported that he put down the paper and said "My God, this is the pandemic. It's here!"[29] Hilleman suspected that such a large number of cases in such a short period of time suggested a highly transmissible agent. A novel influenza virus fit the bill. He also knew that neither his surveillance system nor the WHO influenza surveillance system had detected any new strains circulating. It appeared that they both had missed it. To confirm his suspicions, Hilleman immediately contacted medical personnel in U.S. overseas bases throughout Asia to get him a sample from a person stricken with this new ailment. A medical officer from a unit stationed in Japan tracked down a sailor recently returned from Hong Kong who had taken ill after returning to his unit. The officer had the patient gargle with salt water and spit into a cup. A sample of this material was prepared and shipped to the WRAIR, where Hilleman waited impatiently.

The sample finally arrived on 17 May 1957 and Hilleman immediately injected the material into fertilized chicken eggs to amplify it to produce sufficient quantities to study. He then began to test the strain to determine if it was influenza; if it was a new strain; and whether anyone had immunity to it. The last was demonstrated by mixing the viral material with pooled sera from servicemen and women. The serum drawn from the blood of donors was packed with antibodies which would react visibly to the presence of a virus that the individual had encountered previously. Pooling a large number of sera would generate an imprecise but general summation of whether this type of influenza strain had circulated recently and therefore would be one against which a percentage of the population would have immunity. By 22 May 1957, Hilleman had his answers. The sample was a novel influenza virus; few if any people produced antibodies against it; and, judging by the number of infected in Hong Kong, it appeared to be highly transmissible. In short, it was a pandemic strain of influenza.

Hilleman immediately sent samples of the new virus to the WHO, the United States Public Health Service (USPHS), and the Influenza Commission of the Armed Forces Epidemiological Board (still headed by Thomas Francis) to confirm his findings. And then he took an unusual step. He contacted the heads of the six influenza vaccine manufacturing firms headquartered in the United States and delivered a simple message: don't kill your roosters. As a young boy growing up in rural Montana, Hilleman was aware that farmers sell most of their roosters off for the stew pot at the start of the summer. An influenza pandemic would require lots of new vaccine to protect the public, and producing lots of vaccine required large quantities of fertilized chicken eggs. The roosters would get a reprieve; they still had work to do.

The other health services confirmed the findings of the WRAIR; the strain was new and spreading widely throughout Southeast Asia.

Tests also confirmed that the current vaccine was ineffective against the new strain. It was likely that without a new vaccine, the public would be left as unprotected as they were in 1947, the last time the vaccine had failed. The fear for public health officials was not another 1947 of course, but another 1918. The early reports from Asia did not suggest a revisiting of Spanish flu but a pandemic of the order of the Russian flu of 1889 would be deadly and expensive. A new vaccine would have to be created, fast.

The WHO surveillance system had failed in its task. It had not detected the emergence of a new strain; in fact, it had not even detected the circulating strain when it was at the regional epidemic stage. It had taken a newspaper account to alert the influenza virologists to the new virus. The Organization had not even received a sample of the new virus until late May, when Hilleman had sent one along. After this initial sluggish start, the WHO sped up its response. By 24 May 1957, it was contacting the laboratories in the surveillance network to be alert for the new strain and was already at work creating a seed stock to distribute to vaccine manufacturers around the world.

The United States had not waited for the WHO to create a vaccine template virus. Shortly after delivering the samples he had collected from Japan, Hilleman had sent viral material to the six pharmaceutical firms to begin working it up for production. While manufacturing firms began to craft protective vaccine candidates, the chiefs of the U.S. Military's Preventive Medicine Division met in a hastily called meeting to decide on the armed forces' response, quickly deciding that all military personnel should receive a shot of the as-yet-to-be-created vaccine against Asian flu.[30] U.S.-based vaccine manufacturers now had their first orders.

Interested observers at this military conference to discuss the proper response to the new influenza strains included members of the USPHS. The USPHS wanted to protect its charges too—the public—but it faced different considerations in Eisenhower's America. The military could, and did, order its members to receive the shots. The USPHS had no such powers of compulsion, or at least, did not believe that it did. In addition, whereas the military was ordering inoculations for a stable cohort of relatively healthy young men and women, the USPHS would have to ponder what if any vaccine should be given to an eight- or 88-year-old citizen. Also, the USPHS debated who should purchase the vaccine, who should distribute it, and who should administer the shots. Ultimately, the organization decided to rely upon the private medical system that dominated American medicine, but took on the role of publicizing the need for the injections. Public health was to create the demand; private industry would create the product. The USPHS would also develop a voluntary guideline for prioritizing recipients, but would leave the actual distribution to private industry.

The USPHS also took the opportunity to track and study the virus and detailed the task to the Communicable Disease Center (CDC).[31] Legendary epidemiologist Alexander Langmuir had volunteered the services of the newly created Epidemic Intelligence Service (EIS) for this role. Ostensibly developed for early detection of the use of biological weapons against the United States, the EIS was deployed to track and trace disease outbreaks in initially just the United States, but was soon responding to requests from nations around the world. Langmuir envisioned the service as a type of "medical fire department" for disease emergencies and the new service (it was created during the latter stages of the Korean War) had recently established its *bona fides* by quickly detecting and tracing the accidental infection of polio vaccine recipients whose shots had contained live viral material.[32] Langmuir delegated the task of tracking the spread of the influenza pandemic to his chief assistant, D. A. Henderson (who went on to become a legend himself in the field of public health, directing the WHO program to eradicate smallpox). Henderson recognized that timely information was crucial in tracing a fast-moving infection such as influenza, so he created a system of surveillance reports to announce the appearance of cases and suspect cases of the new strain and distribute these alerts to interested persons both quickly and widely.[33] What Henderson's reports revealed was that the new virus could be detected as early as the opening days of June in a smattering of limited outbreaks throughout the United States.

When Maurice Hilleman had first confirmed that the illness circulating in Asia was due to a novel influenza strain and that this new strain was likely to spread pandemically, he had done more than just alert other medical organizations and pharmaceutical manufacturing firms. He had issued a press release warning the public of an impending epidemic. Further, he boldly predicted that the pandemic would erupt around the United States in September, two to three days after schools opened.[34] Hilleman was specific in his prediction because he saw schools—where children come from a variety of places, are densely crowded together in enclosed rooms, tend to engage in unhygienic practices, and then disperse back into their communities and families at the end of the day—as the perfect nexus for sharing respiratory infections. A September arrival of the pandemic would be disastrous for public health plans for vaccination as the manufacturers would not have had adequate time to produce the sizable quantities of vaccine expected to be needed. Ominously, a mini-epidemic erupted in Louisiana, centered in a region known as Tangipahoa Parish, whose schools opened in July. This region grew strawberries and so the schools closed early in the spring to allow the children to work on the family farms. A steep epidemic attributed to the new strain emerged shortly after the children returned to school. By the time the schools were ordered closed it was too late, and Asian flu spread widely throughout the entire region. USPHS officials feared what would happen when the rest of the nation returned to school in the fall.

Public health officials were right to be concerned. A few days after children returned to class in the first week of September, the new influenza strain rapidly cycled into an epidemic. By the end of the month it was clear that the epidemic curve was on a steep incline and the public clamored for the shots the government had encouraged them to receive. There was little vaccine to be had. Pharmaceutical manufacturers had gone all-out over the summer to produce vaccine, and they had committed to producing at least 60 million doses by 1 February 1958. Unfortunately, they were nowhere near that total by September. Complicating the public's access to the injections was the fact that the military had the first claim on the material produced. Also, distribution to those deemed most needy or most necessary for providing public safety and essential services relied upon voluntary allocation administered by the manufacturers themselves. The result was a disaster where demand greatly outstripped supply and the lack of a centralized distribution system meant uneven dispersal of the protective material and a haphazard order of access to the limited vaccines. Stories circulated widely about a black market for the shots, and citizens muttered about the certain large corporations like Ford motor company and the clerks at Dun & Bradstreet, who had managed to gain access to the shots first. The Asian flu pandemic reached a peak of new cases across the United States by the end of October 1957, largely unaltered by public health measures. As the epidemic wave receded and the number of new cases dropped off, vaccine makers had finally caught up to their production schedules. But by that time, there was little public demand for the inoculations.[35]

In assessing vaccination efforts after the fact, public health officials were forced to conclude that the program had failed to interdict the pandemic in any meaningful way. Only a small percentage of the population had had any protective benefit from the shots by the time the pandemic had arrived in their community. In addition, the distribution of the inoculations was chaotic and uneven and the voluntary allocation plan had failed completely. The experience had also been disappointing for manufacturers, who found that they were saddled with roughly half of the vaccines they had produced; these remained unsold and apparently unsellable.

There were some bright spots amid the rubble of the failed effort, however. Although a little late, surveillance had detected the new strain and correctly predicted its emergence as a pandemic. A protective vaccine had been quickly developed and it had proved to be relatively effective (for those who had received the shot in time). Manufacturers had been able to produce large quantities of vaccine on a crash basis. Scientists had jumped on the opportunity to study a pandemic as it unfolded, following both the virus and the effectiveness of the vaccination programs. Even though they had largely failed in protecting people against the virus, much was learned about it and the transmission patterns of a pandemic. Also, the weakness in the vaccination schemes was starkly exposed; this could provide useful lessons for the next pandemic.

The WHO also took the opportunity to study the epidemic, fostering scientific research into the virus. One study by a Dutch researcher, J. Mulder, suggested that the elderly population had an unusually high antibody reaction to the new strain.[36] This discovery was unexpected and potentially important, since humans generally only produce antibodies to diseases to which they have previously been exposed. Mulder recognized the importance of this finding immediately and realized that he only had a limited time to get corroborating evidence from other geographic regions. Once the epidemic arrived in a community, its high transmissibility meant that so many people would contract the illness that it would be impossible to get a clear picture of antibody rates in the population, especially since some percentage of people would have such a mild course of infection that they may not even know they had been exposed. The WHO quickly distributed Mulder's preliminary findings throughout its network of laboratory associates and his data was soon replicated in the United States and Australia.[37] Retrospective analysis of the course of the epidemic revealed that those aged 70 and above suffered lower than expected infections and mortality rates of Asian flu.

As part of its effort to study pandemic influenza, the WHO issued a call for reports from various states on the steps they had undertaken to respond to the pandemic. The thought was that information gathered from the national pandemic responses to Asian flu would provide useful lessons for future emergency planning. While there is no record of what type of response WHO officials anticipated receiving from their update request, the volume of correspondence that arrived over the next eighteen months must have been a surprise.[38] Big and small states from around the world sent in summaries of their vaccination efforts or requests for seed stock and supplies in crafting vaccine programs. States as small as Iceland and Israel as well as large states such as Egypt and India reported on their vaccine development efforts. WHO officials were surprised at how widespread the technical ability to produce vaccine was; indeed, one unnamed WHO official marked with an exclamation point the announcement by India that it was undertaking vaccine production.[39]

There is, however, a difference between the technical capability to create an effective vaccine and the resources to generate and distribute large volumes of the protective shots. The majority of these vaccines were produced by scientists "at the bench," meaning that laboratory workers themselves injected the material into fertilized chicken eggs, incubated the eggs, harvested the material, purified it, and then distributed it for use. Such a production method is necessarily small-scale. Those who received the shots in a timely fashion were likely protected from the pandemic, but these lucky few were only a small percentage of the population of any state.

Those states like Holland and Australia with the pharmaceutical capacity to produce greater quantities of vaccine—dedicated flocks of chickens, incubation facilities, centrifuge capacity, and bottling and distribution networks—faced the same problems as manufacturers in the United States. It takes time to scale up to the production levels necessary to make a large number of vaccine doses. By the time the manufacturers had reached top production speed, the pandemic was upon them. The vaccine remained in the egg, or in the distribution network, or too recently injected into the patient. Very few people anywhere on the planet experienced any protective benefit from the vaccine campaigns.

The toll for this failure of public health was high. It is estimated that the Asian flu pandemic of 1957 killed 2 million people. The illness itself was not exceptionally deadly, but the virus was so infectious that the sheer number of people infected led to the high death total.[40] Because the availability of vaccine trailed the epidemic peak of the outbreak, the pandemic had risen, crested, and receded largely unaltered by public health.

Public health officials met to ponder what was learned from the pandemic of 1957 and decided that the key variable was time. If vaccine manufacturing could be sped up, if public health organizations were prepared quicker, and if the emerging pandemic could be identified sooner, then it just might be possible to blunt its impact. Influenza researchers concluded that although they might not know when, they were certain that there would be another influenza pandemic. And this time, they were planning on being prepared.

The Hong Kong Flu of 1968

It turns out that the researchers did not have long to wait to test out their new and improved plans, for a pandemic appeared just over a decade later. Events in 1968 were eerily reminiscent of those of the Asian flu pandemic, even down to its initial identification. On 12

July 1968, Charles Cockburn sat in his office reading the *Times* (the *Times* of London, that is).⁴¹ Like Hilleman before him, Cockburn was no ordinary newspaper reader. He was the Chief Medical Officer for Viral Diseases for the WHO and the influenza surveillance system was part of his bailiwick. The *Times* article reported that a number of people in southeastern China were suffering from an influenza-like illness. China lay outside the surveillance network, as the nation had isolated itself under Mao's Cultural Revolution. Cockburn immediately contacted the director of the influenza centre in Hong Kong, a Dr Chang, and asked him to investigate whether there was an upsurge of influenza cases in Hong Kong. On 15 July, Chang replied that a survey of emergency rooms in Hong Kong did not reveal an unexpected number of influenza-like illnesses. But, in a stark reminder of how quickly an influenza epidemic can appear, Chang reversed course just two days later, reporting that presumed influenza cases had risen sharply over the past couple of days. He stated that he was gathering samples and would send them on to the World Influenza Centre in London.⁴² Testing there revealed the illnesses to be caused by a novel influenza strain; it seemed another pandemic was brewing.

The timing of the new virus was unfortunate for public health officials in the northern hemisphere, where the majority of influenza vaccine manufacturers were located. By midsummer most roosters would long since have been fricasseed, so acquiring great stocks of fertilized chicken eggs would be difficult. In addition, vaccine manufacturers would have finished their production of seasonal vaccine and turned their facilities to fabricating other medical supplies.

Nonetheless, vaccine campaigns were ramped up and pharmaceutical houses again began to make viral material on an emergency basis. Public health officials and manufacturers made a concerted effort to move quicker in the various steps of developing seed stock, growing viral material, bottling, certifying for use, and distributing the protective shots, and they were successful. The crash enterprise reached these benchmarks of production more than a month sooner than in 1957. However, it made little difference. As was the case in 1957, the vaccine available to the public lagged behind the pandemic curve. The public in North America clamored for the vaccine when none was accessible in January 1969 and was uninterested in purchasing the shots in February and March, after the pandemic had passed and they had become readily obtainable. This resistance to getting the vaccine shots held true even during the second wave of the pandemic in the fall of 1969, which turned out to be more deadly in Europe and Asia then the first wave.[43]

In a similar fashion to the Asian flu pandemic of 1957, the science and medical community undertook studies of the emerging and rapidly circulating virus, now dubbed Hong Kong flu. Much of their early research on the Hong Kong strain was disseminated through the WHO, which used its journal, the *Bulletin of the World Health Organization*, to publish the papers presented at a special international conference on the pandemic. The conference was jointly sponsored by Emory University, the CDC and the WHO, and a series of papers were presented on the virological aspects of the new strain; the responses of public health in a number of states; reports of various vaccine trials using a variety of approaches, including live virus vaccine studies; narratives on the infection pattern of the strain; and the efficacy of therapeutic treatments. The special triple issue was rushed into print shortly after the October meeting, with the intention of informing ongoing public health protection efforts taken up in response to the new virus.[44]

From the wealth of reports, two types of studies proved to have special relevance for future pandemics. One group dealt with an issue related to surveillance; the other with implications for vaccine production. The surveillance-relevant studies replicated Mulder's antibody studies for 1957. Mulder had found that the aged population had an unusually high antibody response rate to the new Asian flu strain, suggesting that this population had encountered a similar virus previously. Assays undertaken in the United States, Japan, and Holland revealed a similarly high antibody reaction by the elderly population to the new Hong Kong flu strain.[45] These results suggested that the older population had previously encountered it, or a closely related strain. Combined with the data of 1957, the unusual resistance to new strains in the elderly suggested that previous influenza strains return or recycle. And, if this was the pattern, then serological studies of senior citizens could be used predicatively, helping to focus surveillance gathering by detecting to which viral type the elderly had the strongest antibody reactions. For example, if people above a certain age had a strong immunological response to an H1N1 strain (which had not been in circulation for over twenty years by that point), influenza surveillance experts could be on the alert for the reappearance of a new H1N1 strain.

The other important study for future pandemics dealt with vaccine production. Edwin Kilbourne reported his successful attempt to use recombination in generating high-growth/high-yield vaccine seed stock material matched to the novel strain.[46] Scientists were just beginning to understand the issue of viral shift, which, as we have seen, is when a different kind of genetic component is grafted on to an influenza strain type. Researchers had discovered that the outside envelope of the virus was comprised of two elements—hemagglutinin and neuraminidase—and that these signature types of glycoproteins could be replaced with other types of hemagglutinin and neuraminidase. When these new components were incorporated into the virus, it dramatically altered the outside signature of the entity, making antibodies ineffective. The virus had shifted to a new type. This realization of the roles that hemagglutinin and neuraminidase played in typing led to the development of a whole new naming system keyed to these two glycoproteins.[47]

The Asian flu family of viruses was determined to be a virus with hemagglutinin family 2 (H2) combined with neuraminidase family 2 (N2). Hence Asian flu was an H2N2. Hong Kong flu had a hemagglutinin from family 3 (H3) which was combined with the same neuraminidase family 2, so it was an H3N2, representing a shift in the hemagglutinin component of the virus. Kilbourne borrowed this shifting proclivity of the influenza virus and harnessed it for vaccine production. He injected a fertilized chicken egg with a laboratory strain noted for high growth potential (PR8, a member of the H1N1 family of influenza strains) and with the new Hong Kong strain. Kilbourne fully anticipated that the two viral types would swap components, leading to a spectrum of new viruses combining genetic segments from both strains. He then selected the strain with the outside hemagglutinin and neuraminidase characteristics of Hong Kong flu wedded with the fast-growing/high-yielding capacity of the laboratory strain (PR8). The outcome would be a strain that promoted immunity against the new strain (Hong Kong) and generated a high number of copies in a short period of time. The end result would be a much faster-developed seed stock for vaccine use, and one that produced exceedingly well. Recombination technology generated a lot more doses more quickly and so held great promise for public health vaccination planning.

Kilbourne's breakthrough came too late for vaccination programs against the Hong Kong strain, which were already under way when he successfully concluded his experiments. This was unfortunate because after-action studies of vaccine programs in 1968–9 revealed a similar outcome to the efforts of 1957. Few states had the capacity to embark on large vaccination campaigns and those that did found that the availability of the shots trailed the epidemic peak. Distribution systems had also marred the protective efforts, although these failures were not as dramatic as had occurred a decade previously. The toll of the pandemic was lower—an estimated 1 million people succumbed globally—but this had very little to do with medical intervention. Again, public health efforts to protect against an influenza pandemic had failed.[48]

The sweep of the pandemic and the relative failure of vaccine efforts again served to teach scientists and public health experts about the behavior of the influenza virus. The identification of the virus's capacity to swap genetic components clarified how pandemic strains arise, and it could be used to accelerate vaccine production. The fact that the new Hong Kong virus had not been identified until it had become a regional epidemic—much as had occurred in 1957—suggested that the surveillance system needed to be expanded in order to be more effective. Understanding the recycling of pandemic strains would potentially aid this surveillance. The Asian flu of 1957 was an H2N2. The elevated antibody response rate in those aged 70 and older suggested that it was related to the strain these elderly people would have encountered as children; the Russian flu of 1889. Similarly, those 70 and older had high antibody reaction rates to the H3N2 Hong Kong strain, suggesting that they too had been exposed to this type of virus as children. Although no pandemic had been documented 70 years previously, retrospective statistical analysis suggested that the year 1900 was marked by an unusally high number of influenza illnesses. Some influenza scientists held that this elevated year was due to the appearance of an H3 viral type. If Asian flu (H2N2) was a recycling of the Russian flu of 1889, and Hong Kong flu (H3N2) was a recycling of the unnamed epidemic of 1900, the next pandemic on the list was the Spanish flu, which—based on samples recovered from swine later—was presumed to be an H1N1.

A second, surveillance-related theory developed in the early 1970s that was also drawn from studying the history of influenza pandemics. Researchers noted that both 1889 and 1900 and 1957 and 1968 were separated by eleven years. 1900 and 1918 were marked by an eighteen-year gap. Could influenza pandemics appear according to some kind of timing? Public health records revealed that there were sharp epidemics of influenza in 1946 (the origin of the virus that prompted the vaccine failure in 1947) and in 1933 (Andrewes and Smith's year of sneezing ferrets). Perhaps these elevated years of influenza cases were the result of shifted viral types. If this pattern was correct, the next pandemic should arrive somewhere around 1979, eleven years after the appearance of the Hong Kong strain. This was known as the eleven-year-cycle theory.

Finally, in reconstructing the course of events involved in the failed vaccination efforts of 1968, public health officials concluded that vaccination had been ineffectual because the virus had, again, been too speedy for the programs. All steps of the process must be made faster if these guardians of health hoped to protect the public. In the United States this realization that the programs had failed cemented the idea that vaccine distribution and delivery must be centrally managed, which meant by the federal government. The private medical system was unable to fairly allocate the shots. A direct government-controlled system could have a template in place ready to take off as soon as the decision to instigate a vaccine program was made.

All these issues and more were discussed at a series of meetings held in the years following the Hong Kong flu and hosted by the United States National Institutes of Health. The meetings were international affairs, with influenza experts from around the world invited to present their findings and debate the scientific understanding of the virus and how to prevent or mitigate a pandemic. These workshops were subsequently published in the *Journal of Infectious Diseases* for even wider dissemination of the ideas related to influenza. A number of states were prepping for the next appearance of a novel influenza strain, but it seemed that the United States was exceptionally active in this pre-pandemic game planning. u.s. health officials, especially at the CDC, believed that they possessed the manufacturing capacity to generate large quantities of vaccine and the technical expertise to deliver it into the arms of its citizens before the next pandemic crested. In order to achieve this goal, the new strain would need to be identified early, a decision would need to be rapidly made, and the government would have to fully commit enough money and resources to support a large-scale program. It would not be easy, but they believed that it could be done. When the next virus appeared, u.s. officials intended to be ready. As it turns out, they did not have long to wait.

The Swine Flu of 1976

In early January 1976, a mini outbreak of a respiratory disease broke out at a military camp in New Jersey.[49] Subsequent testing determined that the sicknesses were caused by a novel influenza strain and that the infection had been surprisingly widespread in the camp, although the outbreak had died out and no cases outside the camp had been detected. Perhaps most ominously, the new strain was type H1swN1, and testing revealed that it was closely related, but not identical, to swine flu samples from 1930 (gathered by Richard Shope): the presumed descendant of Spanish flu. Both United States health officials and health officials from the WHO were immediately alarmed. The virus was new and had serially transmitted between humans on the base. But the outbreak had stopped and no other cases had been found anywhere around the globe. What was one to make of it? Was it an emerging pandemic or not?

In the United States, mindful of how rapidly the preceding pandemics had outstripped vaccine production, health officials began to lobby for a large-scale vaccination campaign to protect the public. The u.s. had the manufacturing capacity and had been preparing for the next pandemic. The key, they believed, was quick decision-making and getting ahead of an emerging epidemic. Accordingly, in late March 1976, in the absence of any additional evidence, U.S. president Gerald Ford announced that he was requesting that Congress authorize a program to vaccinate every American man, woman, and child. Vaccine manufacturers went into overdrive, seeking to produce around 200 million doses to offset a potential fall and winter pandemic of Swine flu (as the strain was dubbed).

WHO officials, who had also been involved in preparing for a potential influenza pandemic in the years following the appearance of Hong Kong flu, faced a different set of challenges than the United States health officials. The WHO had no vaccine-producing facilities, and it represented a number of states who also lacked the productive resources to mount large-scale vaccine operations. Worse, many of these states had previously purchased the vaccines they did use on a yearly basis from U.S.-based pharmaceutical houses. The U.S. government had now ordered the entire production run of these manufacturers, which meant there would be little vaccine available on the international market. As U.S. health officials accelerated to their decision to mount a national vaccine program, WHO officials began to find the evidence that did not support a pandemic more compelling. Ultimately, the WHO would recommend to its member states that national health programs should continue to monitor the situation, but do little beyond that. In the end, a few states created a small stockpile of protective vaccines, but only Canada undertook serious efforts to develop widespread protective inoculation programs.

When the fall of 1976 came, and with it the traditional onset of influenza season across the northern hemisphere, there was no pandemic of Swine flu. Indeed, there was very little influenza at all. The fall of 1976 was a very mild influenza season. Meanwhile, the United States continued to pursue large-scale vaccination efforts to protect against Swine flu. After overcoming a host of technical and legal roadblocks, the inoculation campaign opened in early October. The vaccine was being produced in bulk, certified rapidly, and distributed efficiently, reaching a peak of 6 million shots injected into the arms of U.S. citizens a week by the end of November. Still, no Swine flu pandemic appeared.

In addition to managing the production, delivery, and injection of the vaccine, the USPHS, through its subsidiary the CDC, tracked the ongoing effort and monitored it for adverse reactions. This focused study was alert for the appearance of potential side-effects in the people receiving shots from this unprecedented health campaign. In early December, a statistical association between being inoculated and the increased likelihood of contracting an obscure neurological affliction known as Guillain-Barre Syndrome (GBS) was detected. In the absence of a Swine flu pandemic, the decision was made to halt the program and more closely examine the relationship between the shots and GBS. The program never restarted. The evidence that is said to support the fact that the Swine flu vaccine actually prompted more cases of GBS remains hotly debated.[50]

The recriminations against the U.S. health officials involved in recommending the program were harsh. The campaign was labeled a fiasco; the money spent, a $135 million waste, with the potential costs even higher once litigation was settled with GBS patients and other claimants. The whole decision-making process was considered rushed by advisors who had panicked. The effort to mount a vaccination campaign against a pandemic that never appeared was considered in some circles an egregious mistake: consequently, blame had to be affixed. The assistant secretary of health in the Department of Health, Education, and Welfare (the department that had overseen the program), Theodore Cooper, and the director of the CDC, David Sencer, were both fired for their roles in the ill-fated program. Meanwhile, the WHO decision for watchful waiting was lauded, although it appeared that there was little understanding of what had motivated that recommendation. The affair was perceived as a black eye for public health.

The Russian Flu of 1977

The influenza season across the northern hemisphere in 1976–77 was comparatively quiet, with only a flare-up of a Hong Kong-drifted descendant known as A/Victoria causing any appreciable level of infection. The onset of flu season in the fall of 1977 was also relatively mild. But in mid-November, reports began to filter through to the influenza surveillance system that there was an upsurge of influenza cases in the far eastern reaches of the Soviet Union. Soviet medical officials sent on a sample of the suspect entity to the NIMR in Mill Hill, London, and later to the Centers for Disease Control (CDC) outside Atlanta. The sample was an influenza strain that came to be called (erroneously, as we shall see) "Russian flu." The infection pattern of the new flu strain was puzzling. Although it was prompting sharply elevated rates of infection, the epidemic struck heaviest on those who were twenty years old and younger. Also, the Russian flu had not driven the previous strain out, which is what generally happened with new strains. Instead, the two strains seemed to be co-circulating. This phenomenon had not been observed before.

The solution to these epidemiological riddles soon appeared. The new strain was an H1N1, although it was not closely related to the Swine flu strain A/New Jersey, also an H1N1. The strain seemed to be most closely related to H1N1's that had circulated before Asian flu had emerged in 1957. In fact, closer study of the Russian flu of 1977 revealed that it was not just *like* a previous strain, it *was* a previous strain: specifically, the "Scandinavian" strain of 1950 that had become a standard laboratory sample.[51] The epidemic, scientists concluded, was likely caused by being accidentally released from a laboratory. Whether that was because of a straight mistake or the result of some type of vaccination test gone awry was unclear. But there was little doubt that the epidemic was man-made.

In tracing back the reports of the new infection, it became clear that while Soviet medical authorities may have been the first to ship samples on to the WHO for identification, the USSR was not the site of the first cases. The new strain was circulating in China from May to October of 1977 and the CDC had independently received a sample of the virus from Hong Kong in November 1977. Much in the same way that WHO officials had been surprised at the number of states embarking on vaccination programs in 1957, WHO influenza experts at the meeting called to recommend responses to the swine flu threat in April 1976 were startled to hear about some of the research programs that had already been undertaken by a variety of states. These officials were most alarmed at the live vaccine studies then under way. The most surprising revelation to these influenza experts was that China had been using live virus vaccines in its research.

Live virus vaccines are an alternate vaccination method from the shots developed by Thomas Francis in 1943. They rely on creating a weakened (attenuated) version of the virus targeted for protection against the illness for which the vaccine was created, and they have a long history in public health. The Sabin oral polio vaccine is an attenuate live polio type, for example. Live influenza virus vaccines have a number of advantages over injectable influenza vaccines. First, they are easier to administer because the vaccine is inhaled and there is no needle involved. Some virologists maintain that the immunity is stronger in those who receive live virus vaccines because the vaccinated receive the actual virus through the method of natural infection. Second, the dosage required is much smaller in a live virus vaccine so greater quantities of vaccine can be produced from each individual chicken egg. There is also less chance of contamination in the preparation, and the manufacturing process is simpler because there are no inactivation steps.

The difficulty in live virus vaccines, and it is a big one, is that the vaccine must be stable. The strain must remain in its weakened state in order to generate protection without causing illness or subsequent spread. Herein lay the difficulty in developing live virus vaccines and the reason why they were not part of the public health armature in the West: researchers in the U.S. and UK found that their developed attenuated vaccines kept getting undermined by the influenza virus's high mutation rate. They could not develop stable vaccines because the virus continually created mutants that were more virulent or more transmissible. Soviet scientists had claimed they had solved these problems and had been using live virus vaccines since at least 1957. But Soviet officials continually declined to provide live virus samples for study by the WHO influenza surveillance laboratories, so these medical researchers were forced to rely upon Soviet claims of safety and efficacy. WHO influenza experts had their doubts about Soviet live virus vaccine attenuation, since they had some evidence of continued transmission of live virus vaccine strains in the Eastern Bloc. These chains of transmission would suggest that the vaccines were not as stable as they were claimed to be.

While there is no clear evidence to suggest that the origins of the Russian flu of 1977 were an experiment gone astray, there is no doubt that the epidemic was man-made. Since 1977, two different types of Influenza A have continued to circulate: H1N1 and H3N2. Although this event did not achieve the same public notoriety as the Swine flu program in the United States, the science and medical community were well aware that someone in the brotherhood had accidentally sparked an epidemic, calling into question their own techniques and practices.

Conclusion

Following the events of 1976 and 1977, feverish preparations for impending influenza pandemics were shelved by public health organizations. The predictive power of the recycling and eleven-year cycle theories was re-examined and abandoned. The notion that a pandemic was imminent receded as new, more exotic afflictions appeared and drew the attention and resources of public health planners. Pandemic influenza began to seem like less of an immediate concern, and to some degree, less of a big deal. The costs to a public health organization of wrongly predicting an influenza pandemic—loss of prestige, authority, and reputation among them —were daunting. The fallout from the U.S. Swine flu program was not lost on anyone in public health.

This is not to say that the field of influenza research had stagnated. The research work on the virus which had begun in the wake of the Spanish flu and sped up following the termination of the Second World War continued, with new breakthroughs helping to unravel the mysterious virus. The surveillance program ground on, continuing to expand its reach by incorporating new laboratories and ratcheting up its ability to detect novel strains. But the model of rapid national vaccination program that had undergirded the responses to Asian, Hong Kong and Swine flu was becoming eclipsed. Previous pandemic response plans had relied on commanding the vaccine produced in a nation-state. But these manufacturers were merging into multinational conglomerates that were headquartered in one nation, producing vaccine in a second, and filling contracts in a third.

Who had the strongest claims on that vaccine was unclear. In addition, the capacity of production closely matched the anticipated market for the vaccines, meaning that there was not much spare capacity to increase production on an emergency basis. Little attention was paid to the fact that the old model for pandemic response was increasingly becoming obsolete and that a new model tethered to new realities did not exist.

The influenza virus operates unconcerned with human affairs. The interest and gaze of the public, and to some extent that of public health as well, may have drifted to other challenges. The virus simply went on with its mindless and endless evolving and swapping of genetic components. The immediate result of this process was that seasonal vaccines needed updating year after year as the prevailing strains drifted beyond the protective shots. But inevitably a strain with radically new components would appear and threaten the world population with a pandemic. Public health focus on this reality of the influenza virus may have wandered during the 1980s and early '90s, but in the late twentieth century both public and professional attention would snap back to influenza pandemics as the virus displayed another one of its endless bag of tricks.

FIVE

Renewed Fears of Flu

The 1980s and '90s were dominated by the appearance of new diseases—most frighteningly, HIV—and the reappearance of those that had been believed to be controlled or on the way to being eradicated. Doubts about the ability of public health to protect its charges began to arise. The buoyant, optimistic attitude surrounding medical advances dissipated, and a somber, somewhat pessimistic view of diseases and health began to emerge. Many people were puzzled and afraid of the sudden appearance and geometric rise of the AIDS pandemic, which, combined with the threat of antibiotic-resistant bacterial infections and the re-emergence of old scourges such as tuberculosis, added to a sense of unease regarding mastery of infectious diseases. Widespread media coverage of these epidemic threats fanned the flame of the public's fear of diseases, hyping, and at times over-hyping, the potential damage of disease transfer. The medical research and public health communities began to craft a new model for conceptualizing microbial dangers to a society.

In an era of heightened public interest and concern with the impact of health threats, two dramatic influenza-related events occurred in 1997. The first was related to the history of influenza, specifically the Spanish flu. The astounding recovery of the deadly virus prompted intense interest in the virus and the devastating pandemic Spanish flu had caused. Its forgotten story was rediscovered, and the appalling mortality it occasioned became a focus of media and scholarly attention alike. The second event was the appearance of a novel influenza strain, and the desperate public health plan to disrupt its transmission and derail a pandemic. The combination of widespread discussion of how terrible an influenza outbreak could be (Spanish flu) and the epidemic potential of a newly emerged strain (Bird flu) brought pandemic influenza back into the forefront of popular attention and public health planning.

The one-two punch of events in 1997 did more than just refocus the public's gaze on influenza: it also prompted an evaluation of the plans in place for public health responses to an influenza pandemic. This examination of public health response models revealed the inadequacies of the programs developed, and forced the realization that the vaccination plans were hopelessly out of date. The global population had soared in the two decades since the last active planning for vaccination in 1976, and vaccine manufacturing had failed to keep pace with it. In addition, largely nation-centered vaccine manufacturing firms had become multinational pharmaceutical conglomerates. Firms had combined, been bought out, or left the market, placing production capacity for protective doses in a small number of international hands. New, effective therapeutic treatments had been developed, the most promising of which in the late 1990s was the drug known as "Tamiflu;" a neuraminidase inhibitor that blocked the action of the neuraminidase glycoprotein and so disrupted the virus's ability to infect new cells. But the stockpile of Tamiflu and other therapeutic courses was woefully small and so would offer little immediate protection. The one bright spot in pandemic planning was that surveillance had continued extending its reach around the globe, bringing new laboratories into the surveillance net and expanding the number of collaborating centers that evaluated the growing number of unidentified samples submitted.

Events in 1997 lent a sense of urgency to the revitalization of pandemic planning. New national and international plans and programs were drawn up, and a reinvigorated effort at coordinating pandemic response programs was initiated. The continual circulation of Bird flu (as the new strain of avian influenza was dubbed) underscored the necessity of not delaying in creating new plans. Although devising a new approach would take time, planning for influenza was back on the docket for public health.

Influenza as an Emerging Disease

The sudden appearance of viral infections such as HIV and the proliferation of antibiotic-resistant strains of bacteria prompted a reassessment of the nature—and future—of disease threats to the human population. The twentieth century had been marked by an apparent steady march of progress in medicine and health that seemed to put infectious diseases on notice that they would soon be controlled or even eradicated. The development of new and highly effective vaccines and the discovery of antibiotics seemed to herald a new age. The reality, of course, was much more complex. While the general public may not have been aware of it, medical professionals understood that antibiotics may be a powerful tool, but are no panacea. Shortly after penicillin was discovered and used, penicillin-resistant strains of bacteria began to emerge. Fortunately, these penicillin-resistant types were susceptible to other, newly developed antibiotics or some multi-drug therapy, but the speed with which bacteria evolved to survive the action of an antibiotic was an ominous portent for the future. None the less, the combination of antibiotics, vaccines, and advances in medical science prompted Nobel Prize-winning scientist Sir Frank Macfarlane Burnet to write—after some careful hedging—that "the most likely forecast about the future of infectious disease is that it will be very dull."[1]

When the "dull" world of infectious disease research heated up in the 1980s and '90s, some members of the scientific community began to rethink the framework for understanding the appearance of these natural agents. A National Institutes of Health (USA)-sponsored conference in 1989 introduced a model that viewed infectious diseases as new or re-emerging threats to society.[2] Vigorously promoted by virologist Stephen S. Morse and Nobel Prize-winning genetic researcher Joshua Lederberg, the paradigm saw the natural world as one where societies needed to guard against the appearance of new, or newly changed, infectious agents. Partly driven by human encroachment into more areas of the planet and by the effects of globalization, which brought the wider world into much tighter connection than at any time in history, this interlinked world facilitated the movement of infections. Human incursions into areas that previously had been thinly populated brought more people in contact with diseases and disease vectors that heretofore had only rarely affected people. Novel infections could also arise because of the swapping of genetic components between microbial entities that create new or substantially different microbial threats. The scale of the effect of epidemic or pandemic spread of these ailments could be quite large. Rather than being merely a danger to individuals or local communities, it began to become apparent that these transmissible elements could serve as a threat to the interests of a state. Thus they required greater attention from national governments.

The driver for this new assessment of infectious disease as a potential threat to national interests was the AIDS crisis. This unknown animal disease had leapt the species barrier and became a transmissible human disease that had quickly mushroomed into a major disruptive killer. The costs of HIV infection in lives, social impact, and monetary costs were significant, and had a major impact on the functioning of governments. In places where the AIDS pandemic was having its most devastating effect, for example in sub-Saharan Africa, the affliction threatened even the continued viability of some governments.

Viewing infectious diseases as threats to the business of the state, rather than just an issue of public health, raised the profile of epidemic disease research and control. Poorly funded public health programs became more lavishly funded from defense budgets. Or, if not directly moved to defense areas, the renewed interest in disease impacts led to increased streams of financing for the scientific research community. Although initially preoccupied with deciphering the dangers posed by dramatic exotic diseases like AIDS or Ebola, the attention of many national governments around the world began to be drawn to the costs of multi-drug resistant strains of bacteria and the re-emergence of afflictions like tuberculosis. Seeking to mitigate the impact of infectious diseases led to increased funding in the scientific study of these infecting agents and to plans to cushion the blow to society if new diseases were to appear. In this emergent disease model for elevated epidemic outbreaks, pandemic influenza was conceptualized as a re-emergent health threat.

The new framework for assessing the potential societal threats from diseases pioneered by Morse and Lederberg received serious attention by the U.S. government, and other national governments also began to envision microbes in this fashion. Throughout the 1990s, this formulation provoked debate and discussion among health and government officials. It was against this backdrop that a revolutionary breakthrough was made in influenza research. In 1997, a scientist from outside the circle of influenza researchers announced that his laboratory had recovered the genetic code of an extinct influenza strain. The identity of this resurrected influenza type was the dreaded Spanish flu. The announcement sent shockwaves through the scientific community.

Recovering Spanish Flu

Jeffrey Taubenberger was not part of the rather small collection of scientists engaged in research on the influenza virus. Instead, Taubenberger led a laboratory located outside Washington, DC, known as the Division of Molecular Pathology at the U.S. Armed Forces Institute of Pathology.[3] The institute housed the National Tissue Repository, which, as its name suggests, stored tissue samples gathered from military physicians. Despite its somewhat lackluster title, the institute was a storage facility for the unusual, interesting, or unknown samples from patients that the military medical corps had collected over the years. It was, in fact, an archive of disease, sickness and injury that stretched all the way back to the U.S. Civil War of 1861–5.

Taubenberger knew this repository was a gem, but it was a gem that few knew about, and he wanted to change that fact. In 1995, there was a lot of talk by the Clinton administration about cutting back government expenses, and Taubenberger did not want his laboratory's funding cut, or even the Institute itself. In weekly staff roundtables, Taubenberger and his staff cast about for a way to show off what the laboratory could do. Some members of staff had recently been working with gene amplification technologies. This technique took fragments of genetic code and made multiple copies (amplified) of the genetic material in a sample. By analyzing (or sorting) these fragments, larger pieces of the genetic code could be stitched together by locating fragments with overlapping sections of the same genetic code. Once the fragments could be identified as the same piece, the portions that extended past these known sections could then be fitted into an ever-increasing string. In this fashion, long stretches of the genetic code can be recovered. Taubenberger and his team thought it might be possible to use this technology to recover the genome of stored samples from their institute and in this way restore the genetic information of a lost disease or affliction. They decided that retrieving the genome of the Spanish flu would be both a scientifically useful venture and one likely to garner favorable attention, proving the value of the Pathology Institute. The first step was to see if the storage facility had stored material from Spanish flu victims.

The process of recovering the influenza strain would not be easy. Finding samples would be easy enough; as it turned out, the Tissue Repository had dozens of specimens from its unfortunate military victims of the Spanish flu. These specimens included virtually every organ and tissue of the body, including the lungs, respiratory tract, and brain. But recovering genetic information from these samples would be difficult. In addition to the technical challenges of using this new process, the samples were almost 80 years old. They had been treated with formaldehyde and then sealed in blocks of paraffin for storage. Any one of these processes could have damaged or destroyed the genetic code and so thwarted recovery of the genome. It was also entirely possible that the samples would not contain any genetic material from the flu strain at all.

In the spring of 1995, the laboratory began the laborious process of searching for slips of the genetic code. The amplification procedure required repeated cycling through different temperatures with a variety of primers that encouraged gene replication. These fragments were then sorted, as scientists looked for the sections of the genetic material that indicated influenza viruses. As is often the case in science, the first efforts at gene amplification were unsuccessful. But the team kept at it. Finally, there was a sample that revealed parts of an influenza strain. As Taubenberger and his team worked through the material, they realized that it resembled influenza virus code, but did not match any known sequenced strain. The fragments they had salvaged were part of the Spanish flu strain.

Recognizing the importance of this work, Taubenberger sent off his preliminary results to the journal *Science*, which published the report in its March 1997 edition. The news came as a bombshell to the influenza research community. Seemingly from out of nowhere and from a laboratory few of them had heard about, the genetic code of Spanish flu had appeared. The stunning breakthrough captured the attention of the popular media. Naturally, after trumpeting Taubenberger's achievement, the popular press began to investigate the dramatic events of the Spanish flu nearly eight decades previously. Few of the general public had been aware of Spanish flu's terrible mortality and morbidity, and so the story generated intense interest.

One of the people captivated by this dramatic discovery of Spanish flu's genetic code was a retired pathologist named Johan Hultin.[4] Like Taubenberger, Hultin too had attempted to recapture the infamous Spanish flu, but Hultin's effort had come more than 45 years earlier. As a medical graduate student, he had proposed an expedition to try and recover viral Spanish flu samples from bodies interred in the permafrost of the Arctic. Although in 1951 the small contingent had successfully located buried victims and recovered tissue samples, they had been unable to recover viable virus. This failed effort rankled Hultin. But as Taubenberger's work demonstrated, viable virus would not be needed. The virus, and its complete genetic code, might lie entombed in the tissues of buried victims. Hultin jumped at the chance to try and recapture the virus again.

Hultin knew that the specimens Taubenberger worked with had been treated with formaldehyde and sealed in paraffin blocks. Either event could have disrupted or altered the genetic code. An intact, untreated, frozen viral sample would eliminate those doubts and questions and provide more material with which to conduct additional tests, as well offering more samples for comparison. Hultin wrote Taubenberger a letter, informing him of his plan to retrieve additional viral material, and set off for the Arctic alone. Returning to the same village he had visited more than four decades previously, Hultin received permission from the village elders for another dig. Among the skeletonized remains buried in the Spanish flu mass grave, Hultin found a woman with a substantial mass of remaining body tissue. Taking samples from the frozen victim, Hultin shipped them to

Taubenberger's laboratory. Remarkably, Taubenberger's team succeeded in recovering the viral genetic material from the tissue. It matched the paraffin-stored specimens almost exactly.

Over the years, using both the stored samples and Hultin's frozen specimens, the Armed Forces Institute of Pathology was able to publish the entire genetic make-up of the virus. The whole dramatic series of events led to much wider interest in the Spanish flu in both public and professional circles, and served to highlight the worst-case outcome of an influenza pandemic.

At the same time that Taubenberger's research breakthrough was appearing in print and shaking up the world of influenza research, a new influenza strain was killing chickens on farms in China. This first appearance of a new influenza strain in the spring of 1997 did not capture much attention outside of a few keenly interested observers, but by the close of the year it would dominate headlines around the world, raising fears of an influenza pandemic. The name "Bird flu" would soon be on the lips of scientists and government planners alike.

The Hong Kong Flu of 1997

In late March 1997, word began to filter back to the Agricultural and Fisheries Department in Hong Kong that a farm in the northwest of the territory had something that was killing a lot of chickens.[5] The outbreak of the mystery illness was dire, with nearly a 100 per cent fatality rate among the 2,000 birds on the farm. Two more farms in the vicinity had epidemics in the subsequent two months, and both farms experienced high chicken die-offs: somewhere around 75 per cent of the birds in both locations. Sudden epidemics among flocks of birds raised for human consumption were unusual, but not unprecedented. Leslie Sims, investigatory veterinary officer of the Agricultural and Fisheries Department in Hong Kong (soon to be promoted to Senior Veterinary Officer in August), was tasked with overall responsibility for veterinary laboratory services in the Special Administrative Region (SAR, as the greater Hong Kong region was referred to as the British prepared to turn the territory over to the People's Republic of China), and began to run tests to uncover the mystery killer.

Sims gathered samples from the dead birds and began a series of surveys to identify the pathogen. A likely candidate for the high chicken mortality rate was Newcastle Disease Virus (NDV), a virulent disease which had produced a high rate of mortality among birds kept in dense concentrations. In fact, it was the successful development of a protective vaccine against this disease that had allowed for the soaring growth of chicken production throughout Asia. Prior to the development and widespread utilization of the vaccine in the 1970s, chicken flocks were necessarily smaller because the virus could rapidly run through a large grouping of the birds. There had been failures of the vaccine in the early years of its utilization, but in the preceding few years the high-quality doses used in Hong Kong had been very effective. Still, it was not impossible that some breakdown in the production of the vaccine or in its utilization had left the birds unprotected from NDV.

Sims's laboratory work quickly ruled out NDV and other possible infectious agents (such as infectious bronchitis). Soon his investigation fingered an influenza virus as the likely culprit. A deadly influenza virus killing the chickens—known in veterinary circles as a Highly Pathogenic Avian Influenza Virus (HPAI)—was again unusual, but not unprecedented. In the agricultural world it was known that influenza epidemics could erupt suddenly with a high degree of mortality in the infected flocks. There had been a number of small outbreaks of "fowl plague" (as it was initially known) and two very sizable ones: Pennsylvania in 1983 and Mexico in 1993. The Pennsylvania case involved the slaughter and death of 17 million birds in order to "stamp out" the deadly infection.[6] Sims himself had been involved in two minor outbreaks in Victoria, Australia.[7] The standard solution to these chicken (and other fowl) influenza epidemics is to rapidly implement biosecurity procedures to isolate the infection in a discrete number of farms, and then destroy and dispose of all the birds on those farms and at surrounding barns a set radius out from the infected flocks. This operation was disruptive and expensive, and had to be done fast, but experience showed that it worked.

Although testing results suggested that the infectious agent recovered from the flocks was influenza, Sims could not be sure, as the samples did not react with any of the influenza reagents he had on hand in his laboratory (a sign, Sims said, of how little attention veterinarians paid to influenza virus in animals). Sims sent some of the samples on to Kennedy Shortridge at Hong Kong University. Shortridge was not only the animal influenza expert in Hong Kong: he was one of the premier researchers of influenza viruses in the world. His particular focus was assaying and tracking the prevalence of influenza strains in animals. After arriving at his post at Hong Kong University in 1972, Shortridge had established a systematic and sustained surveillance of bird populations with an eye to establishing a fuller picture of influenza infections in animal species. Beginning in 1975 and running for the next five or six years, Shortridge sampled an assortment of domestic and wild birds for 51 weeks out of the year (halting only for his staff to celebrate the Lunar New Year). This systematic study—as opposed to spot studies, which only provide a snapshot of circulating viruses—created a rich image of the variety of influenza strains in the wild. Shortridge bolstered his surveillance efforts in bird populations with a systematic survey of pig populations that ran for a couple of years in the 1980s and restarted in the early 1990s. Unfortunately, his sampling efforts were undercut by a persistent scarcity of funds.

Shortridge's research demonstrated that aquatic (ducks and others) and semi-aquatic (geese and others) birds hosted and circulated a variety of influenza strains. These findings supported the theory promoted by William Graham Lever and Robert Webster that aquatic waterfowl are the natural animal reservoir of influenza viruses. Uncovering where influenza viruses came from was an important question, and was one of the research threads pursued by scientists in the decades after the Second World War. It had been determined that a variety of animal species could contract influenza besides humans, including swine, ducks, geese, horses, seals, whales, and a variety of aquatic shore birds. Detecting whether the viruses of these animal species were the same as those of humans, and whether one of these species (or some other unknown species) was the reservoir for influenza, was a research area in which the WHO had been particularly interested. It had funded research into influenza in animals from its earliest days, and in the 1970s this research began to bear fruit. One pioneer in the field was Australian microbiologist William Graeme

Laver, who in his work with longtime collaborator Robert G. Webster demonstrated that all known influenza strains can be recovered from aquatic waterfowl.[8]

Aquatic waterfowl, ducks especially, are the likely home of the influenza virus and in this population it usually causes little harm to the infected carriers. Humans are generally resistant to infection from the myriad influenza strains that circulate in bird populations, for, as we have seen, the receptor binding sites differ in human and bird organisms. But these viruses, or fragments of these viruses, do find a way into the human population, prompting infection, epiemics, and pandemics. Generally, these bird viruses need some sort of adaptive process to infect humans and to become readily transmitted between them.

Humans are not the only species resistant to direct infection from this aquatic waterfowl illness. Shortridge also discovered that only rarely could one detect chickens harboring an influenza virus. Drawing upon his vast library of influenza samples gathered over the years, Shortridge quickly determined that the unknown viral sample sent by Sims was an H5N1 influenza strain, an unusual combination of hemagglutinin and neuraminidase. Shortridge knew that it was rare to ever find an H5 as a component of a strain. He had occasionally recovered H5N2 strains from his bird sampling, but only a handful: one or two a year. By the time Shortridge had typed the virus to H5N1, the bird epidemic had ended. After causing such dramatic mortality on the three farms, the virus had disappeared. Where it came from and where it went were mysteries. Shortridge, however, suspected that this would not be the end of the trouble this H5N1 would cause. The outbreak may have been self-limiting in Hong Kong in the spring of 1997, but somewhere in the wild animal population a virus highly lethal to chickens was still circulating.

Meanwhile, in Kowloon, a highly urbanized part of Hong Kong, a previously healthy three-year-old boy named Lam Hoi-ka developed a fever, sore throat and dry cough on 10 May 1997.[9] Lam's physician diagnosed a throat infection and recommended a course of antibiotics to knock out the invader and aspirin to reduce the fever and lessen the pain. The boy's condition continued to deteriorate and on 13 May 1997 he was taken to a private hospital. Still he continued to worsen, so on 16 May 1997 he was admitted to Queen Elizabeth Hospital in Kowloon. Despite being prescribed a broad spectrum of antibiotics, being placed on a ventilator, and undergoing a battery of tests, the boy died on 21 May 1997. What killed him remained a riddle, and various samples were sent to the Clinical Pathology Unit of the Hong Kong Department of Health. The child's liver showed signs of Reye's syndrome, a complication of an influenza infection prompted by the use of aspirin. Chief pathologist Wilina Lim began to run tests for the influenza virus she suspected as the culprit. These determined that the boy was harboring an influenza A strain, but she could not type the virus because it would not react with any of the reagents she had on hand. Hong Kong was part of the World Influenza Surveillance System, so standard operating procedure required that she send unidentifiable samples of suspected influenza viruses on to the collaborating centers, where unknown strains are typed. Lim dutifully did so, sending on copies of the element to Mill Hill and one to Atlanta. She also sent an additional sample to some Dutch researchers in Rotterdam. The Dutch were interested in studies of the viral drift of influenza strains, so Kennedy Shortridge had facilitated a connection with the Hong Kong Health Department, who would periodically ship samples to Rotterdam.

Even though the strain had been from a fatal case, no great urgency was attached to typing the virus. The freeze-dried sample Lim sent to London turned out to be inactive. Although officials in London should have followed up and requested another sample, they did not. The virus shipped to Atlanta arrived in good shape, but it was not tagged as high priority and was placed with a number of other samples to be typed. Summer is the height of influenza season for the southern hemisphere and so the CDC was busily typing specimens that had been sent from its southern collaborating laboratories. Collaborating centers receive thousands of samples a year, of which the overwhelming majority are drifts of prevailing strains, or ones that should have reacted but did not (gene typing can be a fickle business), and so there is often a delay before unknown strains are identified.

The Dutch researchers, on the other hand, had a much smaller caseload of typing responsibilities, so they were the first to begin the identification procedures. The sample proved difficult to nail down, but by early August lead virologist Jan de Jong was certain of his finding: the virus was type H5N1. De Jong immediately booked a trip to Hong Kong to confirm that the virus was really from the young boy and was not some accidental laboratory contaminant. De Jong's identification of the strain as an H5N1 infection of a human was big news in virology and was a feather in his cap. In the hypercompetitive world of influenza virology, his laboratory had scooped the collaborating centers in Mill Hill and Atlanta (although it should be noted, not Shortridge's laboratory in Hong Kong). The Dutch group got to publish the identification first in *Nature* in October 1997.[10]

The identification of the unusual strain and the fatal case set off an immediate flurry of activity in Hong Kong. Shortridge, on vacation in London at the end of July, recalls returning to Hong Kong and attending a series of meetings with government and health officials.[11] But the activity soon died down; there had been no other cases and, thus far, no other sample of the virus had appeared in the stepped-up surveillance effort. Shortridge, however, remembers being concerned, and returned to his animal surveillance with an even keener eye. Some international health officials also were concerned. Officials at the CDC, in their role as WHO Collaborating Center, scurried to develop reagents to the new strain that could be shipped to the various collaborating laboratories in the surveillance network. If the virus reappeared, they wanted to be sure that it could be identified quickly. But for the most part, influenza and health officials were skeptical that the strain would pop up again. There had only been one case. Complacency over H5N1 was soon to disappear.

On 6 November 1997, another young boy—this one two years old—was admitted to Queen Mary Hospital in Hong Kong with a fever, sore throat, runny nose and cough.[12] Although the course of infection was mild, the child was hospitalized as a precaution because he had a weak heart. The boy recovered and was released, but subsequent testing of samples collected during his illness turned up a positive reaction for H5N1. Alarm bells began to ring. One was a curiosity; two was a pattern. Over the next few weeks another and another person began to be diagnosed as having contracted H5N1. These people were identified after they were hospitalized—the illness caused by the virus was very serious. In the ensuing few weeks, the pace of infected cases began to pick up. By the end of December, eighteen people had been hospitalized with influenza caused by H5N1, of which six were to die (including the original three-year-old boy in May). As the case rates began to tick up in late November, no epidemiological link could be established between the people contracting the virus, but by early December, the epidemiological link became clear.

In the first week of December, chickens in one of the many wet markets then in Hong Kong began to die. Post mortem testing revealed that the cause was H5N1. In short order, positive viral samples began to be recovered from bird cages in a number of markets, including the main wholesale market for the city. Cantonese culture places a high premium on fresh meat. Hong Kong had over 1,000 retail shops and markets that sold, on average, 120,000 poultry—the majority of which were chickens, but also ducks, geese, partridges, quail, pigeons, and a variety of wild caught birds—per day. The markets were a chaotic jumble of cages of a variety of species, including mammals, fish, reptiles, and birds. The buying public wandered through, poking and prodding their potential dinners. In some cases, the buyers took their purchases—flapping and squawking—home for preparation. In others, the animals were slaughtered and gutted on the spot. As Leslie Sims (by this time senior veterinary officer for the Hong Kong Agriculture and Fisheries Department) put it, "when I visited the [Main Wholesale Market] for the first time [in mid-December 1997], it was clear that this was a perfect environment for transmission of disease."[13]

To service this massive market system, a haphazard and free-wheeling interconnected network of farmers, brokers, distributors, and markets arose, with some people playing multiple roles as producers, distributors, and sellers. The farms within the SAR ranged from large operations where the chickens were in very stressful (for the birds) large flocks to small operations where the birds ran free with other farm animals and mingled along the watercourses and ponds with interloping wild birds like ducks and geese. Any and all of these birds—including the wild ones—could end up in the market. This "old style" farming system of close human, bird, and animal interaction that was the pattern in the more remote regions of the SAR was the perfect environment for moving infectious agents from species to species.[14] In addition to this variety of production styles, there developed a series of holding facilities to cushion against the vicis-situdes of the market. These holding facilities, called "chicken inns," held various poultry in temporary conditions depending on demand.[15] The whole system was un- or poorly regulated prior to 1997. As Kennedy Shortridge described it, one of the major markets in Hong Kong was a "temporary" market. The central market was "a temporary market, built in 1972 . . . and it was still temporary in 1997, I think it's still temporary [today]," although it does not get much use.[16]

As large as the supply system was for Hong Kong, the 200 active chicken farms, 68 duck farms, 11 quail farms, and 134 pigeon farms were insufficient to supply the market demands for poultry. In fact, livestock farming was considered a "sunset industry" in Hong Kong, as it was steadily declining. In 1997, Hong Kong farms produced just over 20 per cent of the poultry consumed in Hong Kong. The remaining 80 per cent came from the mainland, and it was this mainland production that was driving Hong Kong farms out of the business. If regulating the motley collection of producers in the Hong Kong SAR was difficult, and a subsequent stringent review of farms and the distribution network revealed a number of problems in maintaining biosecurity, Hong Kong officials had no jurisdiction over mainland farms. Was H5N1 circulating in mainland farms and poultry sheds? Hong Kong public and veterinary officials had their suspicions, but had no way to prove it. The circumstantial evidence was suggestive. The first outbreak of HPAI on the three farms in March–May 1997 occurred in a region close to the border with the People's Republic of China (PRC). Also, the widespread appearance of H5N1 in the variety of fresh markets in Hong Kong suggested recent and continual importation into these sites, but no Hong Kong farms or warehouses had sudden poultry die-offs.

The discovery of the second round of infections set off a mad scramble of influenza experts to Hong Kong. Robert Webster, then laboratory director of the WHO Collaborating Center for Studies on the Ecology of Influenza in Animals, moved into Kennedy Shortridge's facilities at Hong Kong University. CDC representatives headed up a multinational team that moved into the Hong Kong Department of Health laboratories.[17] The brigade of researchers set about a variety of tasks—gathering viral samples from a cross section of markets, farms, and holding pens; tracing the epidemiological contacts of the infected and testing them for recent infection; surveying wild and domestic species for the virus; and drawing blood and serum from a broad selection of people. But by far the most important task that these researchers engaged in was identifying a representative sample of H5N1 that could serve as the seed stock of a protective vaccine.

As we have seen, chicken eggs perform a number of vital roles in both surveillance of the virus and manufacturing protective vaccines. The first step every influenza researcher does upon receiving a sample of the virus is to inject it into fertilized chicken eggs in order to produce an increased volume of material with which to conduct their work. Other uses of chicken eggs included manufacturing reagents to distribute throughout the surveillance network so that laboratories could quickly identify other cases of the novel strain. And, of course, the most important role for the chicken eggs is to make influenza vaccines. Although a number of firms that produced for the European and Asian markets are moving to new technologies utilizing mammal cell line production, for the massive U.S. market only vaccines prepared in chicken eggs are licensed for use. Indeed, at the present time no strain that has been passed though a mammal line at all can be used to produce vaccines for the U.S.

It is for this reason that the H5N1 strain presented a serious problem for influenza virologists. The highly pathogenic H5N1 was not only deadly for chicken flocks: it was equally devastating for their fertilized eggs. Injecting the virus into the chicken eggs killed them in 24 hours. In the eggs, the "chick embryo would be reduced to a gelatinous mess. [A] dark, bloody mess. It would kill the embryo and leave that overnight. You'd inoculate the eggs in the afternoon and next morning it was just a mess. It was such a potent virus."[18]

The virus's destructiveness to chicken eggs presented a serious obstacle to public health officials. In order to be effective, vaccines need to closely match the targeted virus. But the viruses that most closely matched the H5N1 strain killed the production vehicle. Viruses that were not toxic to the egg were poor matches with the deadly flu strain and so afforded only limited immunity. A vaccine seed strain that provided effective immunity protection against H5N1 and could be manufactured in great quantities would have to be discovered or created. This would take time. But the course of events in Hong Kong suggested that time was running out. The number of cases discovered in the human population was increasing each week in December, and it seemed that the H5N1 virus was "undergoing rapid evolution."[19] Thus far, the virus appeared to be a zoonosis, meaning that all the human infections were from people who had come into contact with infected birds, and that the virus lacked the ability for human transmission. Given time, and the virus's high mutation rate, it seemed likely that the virus would develop that ability; and if it did, there would be no chance of stopping a pandemic.

The key to this development of a human transmissible virus was the birds in the markets. By early December it was clear that the poultry in the market were the nexus for human diseases. Once the virus began being recovered from dead birds in the market, they were temporarily closed and a thorough cleaning and disinfecting of the area was ordered. When they reopened, viruses were still being recovered from a variety of places, suggesting that the virus was continuing to be imported and that once in the markets, the virus was amplifying because of the variety and number of birds. By the end of December, fresh imports from the mainland were halted. As one can imagine, as the market/human connection became clear, the buying public stayed away from the market in droves. There was now a large quantity of unsold and apparently unsellable poultry. Shortridge led a surveillance effort of swabbing the respiratory and cloacal tracts of various birds and found an astounding 20 per cent of the chickens were infected with H5N1. This was an astonishingly high burden of viral infection. Chickens almost never host an influenza strain and even sampling ducks and geese had never returned an infection rate higher than 2.5 per cent from any combination of strains. In some cases, the chickens toppled over as Shortridge and his staff walked through the market.[20] It was clear that the birds had to go.

The Hong Kong Health Department—led by Margaret Chan, the current director-general of the WHO—and the international influenza experts were faced with a dilemma. Developing a vaccine would take time, further complicated by the fact that a suitable vaccine strain had yet to be found, and the quickening pace of events suggested that time was running out. The Health Department decided to borrow a page from their veterinary colleagues' book and ordered the destruction and disposal of all the birds in the markets and farms in regions where the virus was detected. The task fell to Leslie Sims in his role as senior veterinary officer for the Agricultural and Fisheries Department in Hong Kong. In order to gather the army of people necessary to slaughter the doomed birds, Sims "dragooned" all government employees to assist in the gassing and proper disposal of the poultry. Following this first-stage eradication procedure, Sims led a complete sanitizing and disinfection program and the establishment of a number of rigid biosecurity programs in the production and supply chain of poultry within the SAR. The immediate eradication program worked. The escalating circulation of the H5N1 was disrupted, and no more human infections with this strain occurred in Hong Kong.

In the wake of the slaughter, Sims initiated a reorganization of the live markets. First, all the live poultry markets were closed for seven weeks and were allowed to reopen only after passing a stringent cleanliness and biosecurity test. The same was true for the farms and distributors that supplied the markets.[21] The markets in Hong Kong itself were also consolidated and reorganized. Markets were separated by species, and also geographically separated. Live chicken markets were located at one end of the city and only serviced by overland truck-rom mainland farms. Live duck sales were banned, as the ducks are the natural host of influenza viruses, and all ducks were delivered by boat to a warehouse on the harbor where they were slaughtered and dressed.[22] The new regulations were only enforceable in the SAR, but mainland farms were strongly encouraged to pursue the same types of changes. To protect the Hong Kong public, Hong Kong health officials tested poultry deliveries from the mainland at the border. Such sampling uncovered an H5N1 strain from a shipment of geese in February 1999, though unfortunately, only after the geese were distributed, sold, and consumed. This recovered virus in 1999 was related to the highly virulent H5N1 strain of 1997, but was substantially different. The discovery did support the idea that H5N1 continued to circulate on the mainland.[23]

The steps taken in Hong Kong and the surrounding areas in 1997 and 1998 were far larger than any previous "stamping out" process—an estimated 100–200 million birds either died or were destroyed in the course of events—and extraordinarily expensive. They were also massively disruptive of people's lives in southeast China and, in the case of Hong Kong, engendered a number of changes in cultural practices. No longer could live ducks be purchased at the markets; instead, only cleaned and dressed ducks were available. The separation and consolidation of the live markets also upended previous lifestyle and shopping practices. But it had apparently been well worth it. The highly pathogenic viruses circulating in November and December were eradicated and no additional infections occurred after the closure of the markets and the mandatory slaughter at the end of December. The actions taken likely disrupted an emerging pandemic: a pandemic of an influenza strain that was highly virulent in humans as well as in chickens.[24]

Retrospective genetic sequencing of this Hong Kong strain of 1997 revealed that this deadly virus was an unusual result of a recombination of a goose strain (H5) joining with two quail strains (H9N2 and H6N1) to generate a new virulent strain known as H5N1. While it had been successful in eradicating the deadly infections of H5N1 in Hong Kong, the slaughter failed to eradicate the source of the novel H5 first uncovered from a goose in 1996. That viral donor has subsequently gone on to spawn a number of H5N1s that circulated widely, creating an unprecedentedly large animal epidemic throughout Asia, Europe, and Africa. Eradication campaigns have been launched in a number of nations, but have only been successful in a handful of them. The virus remains endemic in a variety of wild and domestic bird populations to the present day.[25]

Fortunately, thus far, the virus remains a zoonosis. Human cases have apparently all been caused by direct contact with infected birds. There have been no clearly documented cases of human-to-human transmission. It is this current inability of the virus to be directly transmitted between humans that has prevented its pandemic spread. It is likely that it was the lack of this key ability in the strain discovered in Hong Kong in 1997 that allowed the eradication scheme to succeed. As Kennedy Shortridge observed:

> Certain little observations . . . made me think we've got a
> highly pathogenic virus, but it is not transmissible . . . I don't

think anybody's aware of it . . . [but] there was, in the market, this temporary Hong Kong market . . . five families living there . . . This is the Hong Kong of the older times when everybody sort of lived where they worked. And the market was built around these families. And none of them got sick. You realized in the markets they are talking about, this is the big wholesale market, I looked around, I knew we were in a state of crisis, absolute crisis. Yet when I looked adjacent [to the market] I saw housing blocks, twenty-story housing blocks. None of them got the flu . . . I knew we had a little bit of time on our side.[26]

But there is no guarantee that the virus will continue to be non- or poorly transmitted by humans. The virus has already shown a remarkable proclivity for genetic recombination and the strain has demonstrated a very high mutation rate. Either feature of the virus could result in the creation of a strain that can readily pass between people and thus transit widely.

What, then, did the events in Hong Kong mean for influenza surveillance? Eradicating birds may have become a process to disrupt the chain of infection in poultry barns, but the actions undertaken in Hong Kong were not about protecting chickens. The birds were destroyed and the markets fumigated to protect the human population and hopefully to disrupt the virus's evolution into a human-transmissible and therefore pandemic strain of influenza. This was unusual and a break with previous practices.

Events in Hong Kong rendered traditional vaccination responses to normal influenza infections moot. Vaccination was partly impractical because of the speed of events in November and December, but the fact that the H5N1 virus was lethal to the chicken eggs crucial for vaccine production meant the normal vaccination response had to be changed. Public health experts were forced to improvise. They decided to slaughter all the birds that could conceivably host the virus. To some degree this solution was the natural course of action, since human infections were caused by the victims' contact with infected birds. But the primary reason for the slaughter was because the health officials had no other option. The decision was a desperate measure, taken in the absence of the viability of traditional pandemic response plans.

Here the problems associated with the previous model of public health response to pandemic influenza were brought into stark relief. For a variety of reasons, the situation in Hong Kong was one that could not easily be duplicated. The city had a comparatively contained market system, a sophisticated medical and veterinary infrastructure, a competent and stable local government, and the ability to tap deep reservoirs of funds if needed. Few other states offered the same advantages in the world, especially in Southeast Asia. The public health service had been able to mount a rapid and complex operation successfully. In many ways Hong Kong presented a model that few could duplicate.

The second flaw in the WHO surveillance system revealed by events in Hong Kong was that animal and human surveillance needed to be more firmly integrated. This was not a new observation. As early as 1949, influenza researchers had considered the role of animals in the epidemiology of influenza.[27] WHO and other public health officials had repeatedly stressed the need to incorporate animal surveillance into influenza research. But it appears that this desire for greater integrated research efforts was paid little more than lip service. The two influenza surveillance systems operated largely within their own silos. Nowhere is this separation more visible than in the dual identification of the novel H5N1 virus in 1997. Despite the fact that Kennedy Shortridge's laboratory had identified an H5N1 strain as being responsible for the chicken mortality on the three farms in the SAR by May 1997, nobody contacted his laboratory when an unidentified viral type was recovered from the unfortunate Lam Hoi-ka. Instead, when the Hong Kong Public Health Department was unable to identify the strain responsible for the three-year-old's death, Dr Lim followed WHO Influenza Surveillance standard operating procedure by sending the sample on to the collaborating centers in London and Atlanta. Even when she sought to step outside the system for additional resources, she sought help across the globe rather than across town, sending the specimen to the Dutch laboratory doing human influenza virus research rather than contacting animal influenza researcher Kennedy Shortridge.

The lack of integration between the human and animal influenza surveillance systems at the end of the century is surprising. By this time the theory that aquatic waterfowl represented the natural reservoir for influenza viruses was both well known and accepted. Genetic research on the variety of viral strains had clearly demonstrated that avian influenza genetic components had been spliced into strains that infected humans and that strains that had adapted to human transmission had also been recovered from birds. These facts suggested movement of the virus between species. While it remained true that avian receptor binding sites differed from those of humans, avian type influenza strains had successfully infected humans, as the mounting toll of Bird flu cases demonstrated. The high mutation rate of the virus suggested that if a chance infection of a human occurred, there was a possibility that evolutionary selection pressure might foster the ability for direct human transmission.

A contributing factor for the reticence of human influenza researchers to pay more attention to animal influenza surveillance was the generalized belief that only a limited number of influenza strains were infective for humans. While sixteen different hemagglutinins and nine neuraminidases had been recovered from ducks and geese, only the H1, H2, H3, and N1 and N2 types had previously been found infective for humans. Many researchers evidenced little concern that other types could infect human populations. Hong Kong's H5, soon followed by documented infections of H7 and H9, rendered this hypothesis obsolete.

The events in Hong Kong served as a wake-up call for influenza pandemic protection. The realization by the WHO and a number of national health organizations that they would have been completely unprepared if the events in Hong Kong had been the start of a pandemic galvanized them to examine their plans for an influenza pandemic. Their assessments were not reassuring. In the U.S., for example, revising pandemic preparation had been initiated in 1977, yet twenty years later, the revision had still not got past the study phase. And in this unpreparedness the U.S. was hardly alone.

For the WHO, the events in Hong Kong in 1997 signaled the need to revamp pandemic preparedness in general—both for the organization and for its member states—and to revise the surveillance process. Reliance upon vaccination alone would be unlikely to protect the public from a pandemic of influenza. The virus was too fast and the production and distribution system was too slow to be able to produce enough vaccine to protect more than a small percentage of the world's population. In response to the obvious limitations of a vaccination-only strategy, the WHO created a model that in effect sought to reverse the surveillance process.

As we have seen, the WHO surveillance system was linked to vaccination from its founding. Its structure was a type of hub-and-spoke system that created a hierarchy of select laboratories whose task it was to identify unusual strains. Upon identification, influenza experts would seek to evaluate the potential dangers of the virus and, if necessary, alert the member states that they should begin vaccine campaigns. Such a system carries with it a delay, as the unusual sample is first tested in the laboratories closest to where it has been uncovered. If the sample does not react to the reagents on hand, then it is prepared and shipped to the main regional collaborating center for further testing. The WHO had increased the number to four sites for this advanced testing in order to speed up the process, but the sheer volume of specimens shipped created a significant backlog at each site. Unless tagged as a high priority, the sample could languish at the back of a long queue of material to be tested. This is what occurred in the case of the first strains submitted from Hong Kong at the end of March 1997. Hong Kong demonstrated that a new structure must be created.

The new surveillance system developed in the early twenty-first century created a type of rapid reaction team who would respond when a novel virus was uncovered. This team of experts would rush to the location of the circulating new strain and seek to disrupt its transmission in hopes of preventing a pandemic. In theory the process would work like this: when a new strain was ascertained, the influenza experts would cordon off the infected region. Then, drawing upon the processes that were successful in eradicating smallpox, the health officials would dose the public in a radius around the site with antivirals and experimental vaccines. The idea behind this model drew upon the fact that new viruses transferred to human populations usually take several generations (or passages) to adapt to the new host for efficient transmission. If the virus could be detected early in this process, the adaptation process could be disrupted and a pandemic averted. To further speed up the identification process, health organizations sought to get more powerful virus identification machinery into the hands of these influenza sentinel sites. However the relatively high price of this sophisticated machinery has served as a limiting factor in increasing the power of laboratories in the system.

Such a pandemic interruption program relies upon a robust surveillance system. To that end, the WHO has revamped the International Health Regulations (IHR) with an eye to making it mandatory for all states to report "public health events" that may create an "international public health emergency."[28] H5N1 in Hong Kong also demonstrated the need to pay closer attention to infectious disease events in animal populations. In fact, the appearance of a dangerous epidemic in poultry flocks serves as an early warning signal that heightened surveillance in the human population is necessary. In theory, reports of outbreaks in birds should help focus public health observations to protect humans. Unfortunately, in some cases, where poultry producers fear the economic losses the public health response of slaughter to "stamp out" Bird flu infections may cause them, the surveillance situation has been reversed. Human deaths—which are harder to obscure than chicken deaths—have served as the sentinel for Bird flu outbreaks.[29]

This new surveillance system combines both low- and high-tech approaches to influenza surveillance. Powerful gene-sequencing technologies are employed to identify every letter of an influenza strain. Such tools can be invaluable aids in determining the likelihood of a strain becoming pandemic or how dangerous the strain may be for the public. But this structure also utilizes low-tech solutions. The decidedly non-technical identification of dead birds can be utilized as a sentinel device. Such an approach enables far more states to be actively involved in tracking and tracing potentially dangerous influenza strains. Also, the new approach of disrupting the emerging virus benefits all states, not just the ones that have the means and the resources to mount vaccine programs.

Here, then, is a model that strikes at the root of the problem for the uneven economic and technological gulf that separate wealthy from less well-off states. It was always a conundrum for the WHO surveillance system that identifying an emerging strain relied upon a great number of laboratory sites looking for it. The more sampling and testing sites, the more effective the system. But the states that benefitted from this surveillance were the relative handful that possessed the ability and the resources quickly to produce protective vaccines. Under this new structure, all states benefit from the speed of identifying a new strain, because the plan initiates an operation to interdict the emerging virus before it can gain a foothold in the human population and so develop into a pandemic.

H5N1 in Hong Kong in 1997 was a warning. It was clear to international and national public health experts that a pandemic would have been extraordinarily expensive in terms of both lives and financial costs. The traditional pandemic response to influenza of relying upon the development of a vaccine would most likely have failed. The only possible way to disrupt the chain of infection lay in the somewhat desperate decision of slaughtering all the poultry. Fortunately for everyone, the plan worked. It worked because of the hard work and competence of the medical and veterinary efforts in Hong Kong and the vast resources upon which they could draw. But, truth be told, there was also a fair element of luck involved. The novel strain was uncovered first in Hong Kong, a unique place that had the resources and expertise to identify and quickly respond to the new infection. The medical and veterinary authorities could issue stringent regulations and enforce them rigidly, but their reach did not extend beyond the SAR. Oversight of this process beyond Hong Kong lay with the government of the PRC. In light of the poor performance of the mainland's government when a new animal-transferred virus (known as Severe Acute Respiratory Syndrome or SARS) infected people in wet markets a few years later, one can only marvel that the highly pathogenic H5N1 was successfully eradicated. And in any case, no governmental regulations or stringent rules can control the vast migratory wild bird population that harbors the variety of influenza viruses. Vigilant and heightened surveillance is the key to protection.

The H5N1 flu strain continued to pop up in various nations throughout Southeast Asia during the first decade of the new century, and clusters were uncovered that killed birds in Europe and Africa as well. The roughly 60 per cent mortality rate among those hospitalized with the infection generated high levels of concern among health officials and medical scientists alike. Each human case was scrutinized for evidence that the illness was the result of human-to-human transfer rather than a zoonotic bird-to-human infection. Each bird-killing outbreak was noted and when possible, samples collected. All eyes were focused on Southeast Asia, which was presumed to be the place where Bird flu would evolve into a pandemic strain.

Pandemic influenza was now recast as an emerging disease. Governments recognized that infectious diseases were highly disruptive to a state and could deal a very expensive blow to national and international economies. SARS, which turned out to be a poorly transmitted infection, spread around the world at the cost of at least 774 lives and billions of dollars.[30] Chinese efforts to suppress information about the emerging infection underscored the necessity of transparency in disease reporting. Although it had been circulating at least since November 2002, the Chinese government did not report this "atypical pneumonia" until it began appearing in surrounding states in February 2003.[31] This tightly globalized world required interstate action and coordination in order to protect any particular country's citizens. This truth was especially relevant for fast-moving, highly transmissible infections such as pandemic influenza.

The re-evaluation of the potential dangers and costs of an influenza pandemic set in motion by the events of 1997 re-established influenza as an issue of global and national public health concern. Taubenberger's breakthrough set off a popular and very widely reported discovery of what Spanish flu had done and, by extension, what another influenza pandemic could do. The dramatic appearance and persistence of Bird flu kept surveillance on high alert for the emergence of a human-transmissible version, which would be a likely pandemic strain, since almost no one on the planet has immunity to H5 strains of influenza. The negative assessment of then-current pandemic planning and readiness prompted renewed investment in public health preparation for disasters, especially disasters prompted by the spread of infectious diseases. Surveillance practices were refined and a heightened focus was turned on Southeast Asia. It was, in many ways, reminiscent of 1976. Public health was anticipating the emergence of a pandemic, and believed it knew what type it would be. And when this pandemic emerged, public health organizations were primed to respond very rapidly.

In 1976, the influenza virus had tricked the scientists and medical researchers: a pandemic did not materialize (although admittedly they had consistently stated that a Swine flu pandemic might not appear). In 2009, the influenza virus fooled public health organizations again. A pandemic did appear, but not where the medical researchers had expected and not the strain they had anticipated. Instead of a human transmissible bird flu appearing in Southeast Asia, a swine-origin flu strain (dubbed Swine flu) emerged in Mexico. The best-laid plans of health professionals had gone awry; the virus had stolen a march on public health again.

SIX

Yet Another Surprise

The desperate measures undertaken to head off a potential pandemic of influenza emerging in Hong Kong in 1997 starkly illustrated the limits of protection plans in both national and international contexts. Neither the WHO nor any individual nation-state had a fully articulated pandemic response plan or programs in place to safeguard their citizens. The Hong Kong response in 1997 was a one-off that could not be used as a template for other national responses. Frankly, it had been lucky that the massive bird cull in Hong Kong had even worked.

The potential dangers posed by the H5N1 strain dubbed "Bird flu" galvanized public health planning. As we have seen, health organizations blended high- and low-tech processes to formulate a scheme designed to disrupt a pandemic strain before it had fully adapted to the human host. Although never explicitly stated as such, the WHO pandemic response plan was really designed to interrupt the adaptation of Bird flu to human populations. Since the avian virus lacked the ability to be spread by humans, and it was believed that only transmission through a series of people would adapt the virus to readily infect this population, the plan hinged on the idea of catching this Bird flu before it fully adapted for person-to-person transference. The success of this interdiction effort relied upon a robust surveillance system. The then current surveillance system, while efficient in some ways, was far from perfect, though it would have to be near perfect to detect an emerging human Bird flu strain.

As part of this effort to improve the effectiveness of influenza surveillance, U.S. researchers were testing some enhanced survey techniques around San Diego.[1] On 30 March 2009, a ten-year-old boy took sick with a fever, cough, and vomiting. The boy was taken to an outpatient clinic in San Diego County, where his symptoms were treated and he subsequently recovered. A swab from his nose and throat was taken and sent on for further testing as part of a clinical study in identifying the sources of influenza-like illnesses. Exams in the laboratory revealed the sample to be an Influenza A strain, but the specimen did not react to the reagents on hand for H1N1, H3N2, or H5N1—the two circulating seasonal strains and the feared Bird flu. The unidentified influenza sample was shipped on to the CDC in Atlanta for identification. It arrived on 14 April 2009. Meanwhile, another sample of an unknown influenza virus was also winding its way to Atlanta. A nine-year-old girl in the county adjacent to San Diego had contracted a high fever and cough, whose onset began on 28 March 2009. She was taken to a naval health research center for treatment. A testing sample was also collected from her as part of a clinical study on influenza surveillance, which revealed the specimen to be an unknown influenza type. This sample too was sent to Atlanta for identification, arriving at the CDC on 17 April 2009.

The hyper-alert influenza reference section of the CDC quickly recognized the importance of these two specimens. Testing and a quick epidemiological follow-up revealed some pertinent facts. First, the virus was a type of swine influenza strain. Second, the two children had no known epidemiological link. Third, the children had no known interaction with pigs. These facts suggested that the children had contracted the swine influenza in some other fashion, most likely by human transmission. It appeared that Swine-Origin Influenza A (H1N1), or Swine flu, was a new strain that could be passed from person to person.

The Swine Flu of 2009

Events related to the Swine flu of 2009 unfolded very rapidly. As soon as the CDC determined that the viral strain from the two cases was a swine flu, they immediately broadcasted this information via the *Morbidity and Mortality Weekly Report* (MMWR). The MMWR (and the CDC's WHO counterpart, the *Weekly Epidemiological Record*) is a clearing house for the interesting and useful epidemiological reports from around the world collected by the CDC. The reports are short and data-rich and are widely distributed to medical and public health professionals globally. The intent of these brief reports is to inform physicians and scientists about infection patterns quickly so that they can be alert to the appearance of these unusual ailments or conditions in their own districts and clinics. In response to the discovery of Swine flu, the CDC backed up this informal method of calling for enhanced surveillance by formally requesting that county health departments in Southern California immediately increase the number of samples collected to test for possible influenza infections. The organization also requested that any specimens that could not be immediately identified be sent to Atlanta without delay.

The CDC requested greater numbers of assays be conducted because, in general, only a small percentage of illnesses that present symptoms typical of influenza are ever definitively tested to ascertain whether a patient is actually harboring the virus. During influenza season, so many people appear in medical clinics and doctors' offices with fevers, coughs, and other typical influenza symptoms that doctors generally assume they have the flu and issue treatment accordingly. It is for this reason that it was fortuitous that these first two children had turned up at clinics involved in testing enhanced surveillance procedures. If they had been taken to an outpatient facility not involved in the study, or even these two facilities at another time, it is unlikely that these swine influenza infections would have been uncovered when they were.

The CDC request for increased disease detection bore dividends almost immediately. Samples flowed in from Southern California, revealing additional cases, as well as cases uncovered in Texas. The epidemiological link between these two regions became clear when it was reported that on 12 April 2009, Mexican authorities had informed the WHO (per new international health regulations) that there were a multitude of clusters of unidentified influenza-like illnesses circulating in several of their states. Specimens sent to the CDC from Mexico confirmed the suspicion that the sickness in these Mexican citizens was from the same strain of Swine flu that had been recovered in San Diego. Moreover, additional confirmed and suspect cases were being found in more states in the U.S., including

New York, Illinois, and Arizona. Very quickly, other nations around the world began to report suspected cases of the new flu as well. An epidemic was under way.

The rapid explosion of the new Swine flu infections thwarted international public health plans to disrupt a pandemic. The virus was the wrong type: Swine flu, not Bird flu; had appeared in the wrong place: North America, not Southeast Asia; and appeared to be already transmissible between humans. As part of their influenza pandemic alert system, the WHO had created a chart with levels informing health professionals what the WHO assessed as the current likelihood of a pandemic.[2] The stages ranged from 1 (no animal influenza circulating) to 6 (community-level outbreak involving two or more WHO regions). In the spring of 2009, the WHO gauged the pandemic alert level to be at phase 3 (small clusters of an animal influenza strain that infects a limited number of people, but no sustained human transmission) because of the circulation of Bird flu. Tied to these phases were recommendations from the WHO regarding pandemic protection activities that their member states should be beginning or planning. The levels also triggered WHO activities such as pandemic interdiction efforts and the release for use of stockpiles of therapeutic medicants such as Tamiflu and Relenza. The WHO had been purchasing courses of these neuraminidase inhibitors for use in regions where an influenza strain first emerged. The fast-moving infection forced the WHO to raise its pandemic alert level to 6 by early June 2009, which meant it had determined that a pandemic was currently under way.

The speed of the 2009 Swine flu strain rendered WHO planning moot. There was no chance to interrupt the pandemic transmission of the virus, and the stockpiles of drugs to protect citizens were paltry considering the massive numbers of people at risk of infection. The only option left for public health was the old standby—rapid manufacture and distribution of a protective vaccine.

There were some minor differences in the programs of vaccination undertaken in response to Swine flu compared to those of previous crash immunization programs. For example, although the U.S. still mandated that all vaccines be manufactured using the chicken-egg production model, a number of European and Asian states had moved to the faster cell-line fabrication methods. Also, unlike in previous campaigns, there were useful medical treatments that protected people or shortened the course of infection. But the challenges faced by public health practitioners in the twenty-first century would be well understood by their twentieth-century forebears: the fast-moving virus quickly outstripped production and distribution of effective treatments. As had been the case in both 1957 and 1968, the pandemic spread, peaked, and declined largely unaltered by protective actions from health organizations.

Assessing the Swine Flu of 2009

As the Swine flu pandemic spooled out during the summer and fall of 2009, it soon became clear that this was no Spanish flu redux. While the virus was successful in transiting widely and rapidly, the illness it occasioned was comparatively mild. Fears that a surge of sickened influenza patients would overwhelm healthcare facilities were not realized. More often, the anterooms of physicians and the wards of emergency rooms were filled with people fearful of contracting Swine flu, rather than patients in dire need of treatment. Shortages and delays in acquiring vaccine produced some tension, but for the most part hysteria over a pandemic remained muted. Initial reports suggested that the affliction was prompting relatively low levels of mortality, perhaps even lower than the number a seasonal visitation of influenza reaps.

Because of the apparent mildness of the Swine flu outbreak, charges began to circulate that predictions of the potential impacts of a pandemic had been purposely inflated. Much of this type of criticism focused on the activities and proclamations of the WHO, which had been in the forefront of recommending responses to the new strain. When the WHO elevated its assessment of Swine flu to level 6 in early June, it triggered the release of therapeutic stockpiles and medicines it controlled, as well as urging its member states to begin producing vaccines. Some people charged that the WHO recommendations were driven by the desire to inflate the bottom lines of pharmaceutical houses rather than because Swine flu represented a real public health threat. Initially, these charges were limited to the fringe media and blogging sites, but these accusations began to gain credence when they were discussed in the mainstream media. The criticisms of WHO motives reached a head when an investigative report published in the *British Medical Journal* revealed that a number of influenza experts were receiving monies from pharmaceutical firms for research and spokesperson roles. These experts had previously helped craft the WHO pandemic response plans and were also advising the director-general on what the WHO should recommend to member states in response to the then-current Swine flu crisis. They were also the group that helped determine what phase of alert should be triggered by the spreading virus. This apparent conflict of interest, and the vocal denunciations of those involved, prompted WHO director-general Margaret Chan to issue a public reply and to order an independent study of the process by a committee headed by an independent outsider.[3]

The vehemence of the attacks against the development of large-scale vaccination efforts and distribution of therapeutic antiviral courses seemed to stem from the fact that the pandemic was not as dire as the worst-case scenarios had predicted. As evidence for this charge that threat of Swine flu was purposely hyped for nefarious reasons, critics pointed out that the mortality toll of this strain was comparable to that of an ordinary influenza year. Seasonal outbreaks of the virus do not require massive government-supported inoculation efforts, nor purchase and subsidy of expensive antiviral treatments. Therefore, this line of thinking runs, the extraordinary efforts were unnecessary and the funds allocated were wasted.

Whether the money was wasted—or, to state it more formally, whether the risks to society posed by the influenza virus warranted the expenditure of funds—relies upon a value judgment of affixing a price to health and life. But making this type of calculation has to wait until a final accounting of Swine flu can be totted up, and in this case the critics were premature in their judgment. The assertion that the pandemic was mild and comparable to a seasonal outbreak of influenza was drawn from the laboratory-diagnosed mortality figures of the pandemic. Relying upon these numbers has two sets of interrelated problems, however. First, only a small percentage of influenza cases are definitively tested to determine if the illness is prompted by the virus. And, as was suggested by a three-state study undertaken in the U.S., even in cases where physicians were looking for influenza as the cause of death, a number of cases were accidentally overlooked. In fact, in this U.S. study, when more powerful genetic testing technology was used, 20 per cent of unexplained deaths were subsequently discovered to have been related to Swine flu.[4] In addition, as we have seen, influenza causes elevated death rates across a number of categories, so the only way to measure a pandemic's mortality impact is to count up the number of extra dead. The total accounting of the public health figures drawn from this data has yet to be collected and thus the full analysis of the pandemic remains incomplete.

The second problem of relying upon laboratory-observed Swine flu infections is that the critics are equating the mortality of a seasonal year directly with the numbers estimated for the Swine flu pandemic year. On one level that comparison is correct; a victim who dies from pandemic influenza or from a seasonal case still counts as one. But, drawing upon another method of calculating the costs of a death— a calculation based upon the number of Years of Life Lost (YLL)—the Swine flu pandemic can be seen as having a much higher social cost. Using this metric, the death of a young person has a greater number of YLL than if an aged person dies from influenza. Assuming 75 as the average life span, a grandmother who dies at age 70 has five YLL from her untimely death; whereas if her twenty-year-old granddaughter succumbs to infection, she has 55 YLL. Both are a tragedy, but the loss of the younger citizen means that a greater number of productive years she could have contributed to society is lost. As discussed previously, the typical victim who succumbs to an influenza infection and its complications is either very young or old. But the Swine flu of 2009 was not typical: the mortality blow struck heaviest on the younger population. Indeed, the mean age of death recorded in the Swine flu pandemic in the United States was 37.4; almost four decades younger than the average mean for seasonal influenza for the years 1979– 2001 (75.7 years). Swine flu's mean age of mortality was second only to that of Spanish flu, whose mean age of victim was a dreadful 27.2 years. By this measure, Swine flu's impact on society was significantly greater than that of a normal seasonal year of influenza, even if the final tally of Swine flu is less than the yearly average of influenza.[5]

Conclusion

The Swine flu pandemic of 2009 is merely the latest chapter in the relationship between the influenza virus and the human population, and it nicely illustrates both what is old and new about the viral entity. The virus possesses the ability to move between species, shifting and combining to generate new versions and prompt widespread human infection. The Swine flu of 2009 is a perfect example of this phenomenon, as that particular strain was a triple assortment that contained gene segments from human, avian, and two distinct swine lineages (North American and Eurasian) combined together into one package.[6] The virus's ability to circulate through a variety of species before its emergence as a human pandemic strain is remarkable and highlights the necessity for integrated multi-species surveillance of influenza strains. In the case of the Swine flu of 2009, phylogenetic evolutionary analysis suggests that the virus had been circulating in swine populations for more than a decade before its final combination, and the triple-reassortant strain had entered the human population several months before it was detected as an emerging pandemic.[7] The capability of the virus to move between species is age-old, but our capacity to detect and track this movement is new. To take advantage of this new technological power, animal and human disease tracking should be much more firmly integrated. The more we study and understand human diseases, the more apparent it is that we are firmly linked to our animal cousins and the environment we share.[8]

While the movement between species of infectious diseases such as influenza may be a very ancient process, the speed with which these afflictions circulate, as well as the number of potential hosts for the infection, has sharply increased for both human and animal populations. Humans have increasingly become an urban-centered species. The steep population expansion in the late twentieth and twenty-first centuries has been reflected in the rapid growth of megacities (of 10 million people or more) that are both tightly linked to their hinterlands through the continual movement of people and goods and wedded to other megacities around the globe. In addition to this expansive growth of human populations crowded together, modern industrial farming practices have created mega-populations of domestic animals—especially cattle, swine, and fowl—that, like humans, are densely packed into small spaces. Rather than ranging dispersed in large enclosed areas, these animals destined for the dinner table are crowded together in feed lots or stacked high in barns and sheds. This economy-of-scale production system has also fostered a large and increasing international distribution network where animals are born or hatched in one location and shipped elsewhere for fattening and final processing. In addition to providing increased quantities of cheaper meat, the process also inherently serves as an excellent distribution system for animal pathogens.

Some have charged that this unnaturally dense packing of herds and flocks has created an environment perfect for generating highly transmissible infections, and ones whose increasing levels of virulence are not evolutionary dead ends.[9] In other words, because of the artificially close proximity of new susceptible organisms, highly virulent agents, which in the wild would be weeded out by natural selection, are allowed to continue their chain of transmission, even if they rapidly kill their host. Indeed, such a system would present an environment that would favor faster-transmitting organisms regardless of the damage to the host, because the ability to first infect a new organism would be more evolutionarily highly prized than a longer time period of infection.

The Swine flu of 2009 also illustrates the maddeningly unpredictable nature of the influenza virus. While observers were straining to detect the first movement of Bird flu into the human population in Southeast Asia, Swine flu emerged and circulated on the other side of the globe. Though its relative Spanish flu killed many millions in 1918, the H1N1 Swine flu was significantly milder in impact (even though the final tally still waits totaling up). The only prediction one can make about influenza and influenza pandemics is that they will remain unpredictable.

Still, the long history of influenza pandemics allows us to make some modest assertions. The influenza virus will remain a pandemic threat to human populations for the immediate future. The illness moves too fast for current vaccine production and distribution technologies to intercept, and its rapid rate of evolution holds out the constant threat that presently effective therapeutics will be rendered useless. Therefore, both expanding the surveillance web—including much more integration of animal disease surveillance—and increasing the speed of detection remain important aspects of pandemic preparedness. We do not know whether early detection of an emerging influenza strain will enable interdiction efforts to succeed, but, at the very least, the sooner a new virus is identified, the sooner protective vaccines can be created and delivered to safeguard the public. As we have seen, influenza pandemics are no mean threat. They are expensive in both lives and financial costs, and protecting citizens from this illness through the use of vaccines and medicants is both cost-efficient and the right thing to do. The influenza virus kills millions every year and pandemic strains strike heavily across all age groups, amplifying the costs of infection.

Finally, it is important to keep a sense of perspective in mind as well. I have argued throughout this book that influenza pandemics are important and a serious threat to health. They are deadly and disruptive to society and economics. But they are not apocalyptic. More than 97 per cent of the people infected by the worst-ever pandemic recorded recovered, and even at the height of the pandemic, more people were without the flu than with it. Hysteria has no place in discussing public health preparations, and informing the public about potential risks should not tip over into scaremongering. However, finding the balance, where the public is suitably alert to the threats of a pandemic without being panicked by worst-case scenarios, remains a challenge for evaluating all infectious diseases, not just influenza.

The influenza virus continues its mindless combination and mutation processes, and there is suggestive evidence that changes induced by industrial farming and international commerce may be accelerating the generation of new strains. Science has made great strides in unraveling the mysteries of influenza as well as providing new and effective protection options. While controlling influenza does not appear to be likely in the near future, the powerful new detection tools and program modeling offer public health a real chance to mitigate a pandemic's impact. The challenge, however, is that each time it appears that science has figured out its secrets, the virus pulls a new trick out of its sleeve. We may have gained a great amount of knowledge about influenza in the last few generations—knowledge that would stagger our ancestors—but, as the Swine flu of 2009 has reminded us yet again, we still have much to learn.

References

Introduction

1 Transcript of Statement by Margaret Chan, Director-General of the World Health Organization, 11 June 2009, available at www.who.int/mediacentre, accessed 10 June 2011.

2 The nomenclature for influenza will be discussed in chapter One.

3 Such an accounting does not include those with mild or unnoticeable infections of Bird flu so the real rate is certain to be lower. Still, the very serious and often deadly course of infection with Bird flu is alarming. See "Cumulative Number of Confirmed Human Cases of Avian Influenza A/(H5N1), reported to WHO, 2003–2012," available at www.who.int, accessed 25 June 2012.

4 "Swine Influenza A (H1N1) Infection in Two Children-Southern California, March–April 2009," *Morbidity and Mortality Weekly Report*, LVIII/15 (24 April 2009), pp. 400–02.

5 Governments worldwide had about 220 million doses of antivirals in stock at the time of Chan's announcement. The WHO had reserved enough Tamiflu to treat 5 million people. "Graphic: Antiviral Stockpiles," Reuters.com (30 April 2009), available at http://blogs.reuters.com, accessed 13 June 2011.

6 For example, in the United States 40 million doses—nearly a quarter of the ordered production—were never used and ended up being destroyed. See Mike Stobbe, "Millions of Vaccine Doses to be Burned," for the Associated Press in *The Virginia Pilot* (2 July 2009), p. A1, accessed via LexisNexis Academic, 13 June 2011; see also "40m Doses of H1N1 Vaccine to be Destroyed," UPI.com, available at www.upi.com, accessed 13 June 2011.

7 The CDC estimates that there were about 12,500 deaths in the U.S. attributed to 2009 H1N1 influenza as compared to 36,000 flu-related deaths per year on average. See "2009 H1N1 Flu," www.cdc.gov, accessed 13 June 2011. The WHO reported (as of 6 June 2010) over 18,000 laboratory-confirmed deaths from 2009 Swine flu from 214 countries. While certainly a vast undercounting of the pandemic's real toll, the low numbers do provide indications that the pandemic was mild. See "Global Advisory Committee on Vaccine Safety, 16–17 June 2010," *Weekly Epidemiological Record*, LXXXV/30 (23 July 2010), pp. 285–8.

8 Mark Honigsbaum, "Was Swine Flu Ever a Real Threat?," Telegraph.co.uk (2 February 2010), accessed 18 June 2012; Imogen Foulkes, "WHO Faces Questions over Swine Flu Policy," BBC News Europe (20 May 2010), available at www.bbc.co.uk/news, accessed 18 June 2012.

9 Symptoms from CDC website, "Flu Symptoms and Severity," available at www.cdc.gov/flu, accessed 15 June 2011.

10 For names see Charles Creighton, *A History of Epidemics in Britain*, vol. II: *From the Extinction of Plague to the Present Time* [1891] (New York, 1965), pp. 304–14, and John F. Townsend, "History of Influenza Epidemics," *Annals of Medical History*, n.s. V/6 (November 1933), pp. 539–42.

11 Aside from the dramatic Spanish flu pandemic, the average case fatality rate for an influenza infection, even in a pandemic year, is <1 per cent. See Jeffrey K. Taubenberger and David M. Morens, "1918 Influenza: The Mother of All Pandemics," *Emerging Infectious Diseases*, XII/1 (January 2006), p. 15.

12 See the 1803 Temple West illustration "An Address of Thanks from vthe Faculty to the Right Hon. Mr Influenzy for his Kind Visit to the Country," at http://medphoto.wellcome.ac.uk, image L0009997. Influenza, with its proclivity to facilitate secondary pneumonia infections, prompts elevated mortality among the elderly. See David S. Fedson, Andre Waida, J. Patrick Nicol, and Leslie L. Roos, "'The Old Man's Friend,'" *The Lancet*, CCCXLII/8870 (28 August 1993), p. 561.

13 For CDC estimates see "Prevention and Control of Influenza with Vaccines," Recommendations of the Advisory Committee on Immunization Practices (ACIP), 2010, available at www.cdc.gov,

accessed 13 June 2011; for estimated costs see Noelle-Angelique
M. Molinari et al., "The Annual Impact of Seasonal Influenza in the
U.S.: Measuring Disease Burden and Costs," *Vaccine*, XXV (2007),
pp. 5086–96; for global estimates of mortality see Scott P. Layne,
"Human Influenza Surveillance: The Demand to Expand," *Emerging
Infectious Diseases*, XII/4 (April 2006), p. 562.

14 The genetic elements that underlay seasonal and pandemic influenza
will be discussed in chapter One.

ONE
Know Your Enemy

1 The following discussion of the virus is drawn from L. R. Haaheim,
"Basic Influenza Virology and Immunology," in *Introduction to Pandemic
Influenza*, ed. Jonathan Van-Tam and Chloe Sellwood (Oxford, 2010),
pp. 14–27; Karl G. Nicholson, "Human Influenza," in *Textbook of
Influenza*, ed. Karl G. Nicholson, Robert G. Webster, and Alan J. Hay
(Oxford, 1998), pp. 219–64; and Edwin Kilbourne, *Influenza* (New York
and London, 1987), pp. 25–56.

2 Unless specifically noted otherwise, the following description of the
human immune system is drawn from Jan C. Wilschut, Janet E.
McElhaney, and Abraham M. Palache, *Influenza*, 2nd edn (Edinburgh,
2006); Darla J. Wise and Gordon R. Carter, *Immunology: A Comprehensive
Review* (Ames, IA, 2002); Mary S. Leffell, Albert D. Donnenberg, and
Noel R. Rose, eds, *Handbook of Human Immunology* (Boca Raton, FL,
1997); and James A. Marsh and Marion D. Kendall, eds, *The Physiology
of Immunity* (Boca Raton, FL, 1996).

3 Replication estimates from Claude Hannoun, keynote lecture at "After
1918: History and Politics of Influenza in the 20th and 21st Centuries,"
Rennes, France, 25 August 2011.

4 As Shanks and Pyles describe it, this immunity production process is
a very Darwinian survival of the fittest model. The antibody with the
best match is stimulated to produce more, and because the production
of this "best match" has variation, it eventually results in an antibody
that is a perfect fit for the target. See Niall Shanks and Rebecca A.
Pyles, "Evolution and Medicine: The Long Reach of 'Dr Darwin',"
Philosophy, Ethics, and Humanities in Medicine, II/4 (2007).

5 For the definitive account of smallpox's eradication see F. Fenner, D. A. Henderson, I. Arita, Z. Jezek, and I. D. Ladnyi, *Smallpox and its Eradication* (Geneva, 1988).

6 The following discussion of the genetics of the influenza virus is drawn from Kilbourne, *Influenza*, pp. 111–56; Nicholson, Webster, and Hay, eds, *Textbook of Influenza*; Washington C. Winn Jr, "Influenza and Parainfluenza Viruses," in *Pathology of Infectious Diseases*, vol. 1, ed. Daniel H. Connor, Francis W. Chandler, David A. Schwartz, Herbert J. Manz, and Ernest E. Lack (Stamford, CT, 1997), pp. 221–7; and Haaheim, "Basic Influenza Virology and Immunology," pp. 14–27.

7 John Holland, Katherine Spindler, Frank Horodyski, Elizabeth Grabau, Stuart Nichol, and Scott VandePol, "Rapid Evolution of RNA Genomes," *Science*, n.s. CCXV/4540 (26 March 1982), pp. 1577–85.

8 The connection between avian waterfowl and influenza viruses was first proposed by Graeme Laver and Robert Webster in a series of studies in the late 1960s and early '70s. For an overview of the people and processes involved in developing this theory of an avian home of influenza see William Graeme Laver, "The Origin and Control of Pandemic Influenza," *Perspectives in Biology and Medicine*, XLIII/2 (Winter 2000), pp. 173–92. For a commemoration of Graeme Laver see Robert G. Webster, "William Graeme Laver," *Biographical Memoirs of Fellows of the Royal Society*, available at http://rsbm.royalsocietypublishing.org, accessed 5 June 2011.

9 It should be noted that some mutants of the H5N1 strain circulating in Eurasia since 1997 (colloquially called Bird flu), has been found to cause levels of mortality in duck and geese populations from time to time.

10 See R. G. Webster, K. F. Shortridge, and Y. Kawaoka, "Influenza: Interspecies Transmission and Emergence of New Pandemics," *FEMS Immunology and Medical Microbiology*, XVIII (1997), pp. 275–9.

11 Homo Sapiens also have some sialic acid alpha 2,3 receptors, but they are located deep in the lungs making both infection of these cells difficult because of their location and transmission of the virus to other hosts problematic because it takes an unusually deep cough to expel the virus. It is partly for this reason that people who contract Bird flu have thus far been unable to infect those around them.

12 C. Scholtissek, "Pigs as the 'Mixing Vessel' for the Creation of New

Pandemic Influenza A Viruses," *Medical Principles and Practice*, 11/2 (1990/91), pp. 65–71.

13 Mikhail Matrosovich, Alexander Tuzikov, Nikolai Bovin, Alexandra Gambaryan, Alexander Klimov, Maria R. Castrucci, Isabella Donatelli, and Yoshihiro Kawaoka, "Early Alteration of the Receptor-Binding Properties of H1, H2, and H3 Avian Influenza Virus Hemagglutinins After Their Introduction into Mammals," *Journal of Virology*, LXXIV/18 (September 2000), pp. 8502–12. For the Swine flu of 2009 see Rebecca Garten et al., "Antigenic and Genetic Characteristics of Swine-Origin 2009 A (H1N1) Influenza Viruses Circulating in Humans," *Science*, CCCXXV (10 July 2009), pp. 197–201.

14 Richard E. Neustadt and Harvey V. Fineberg, *The Swine Flu Affair: Decision-making on a Slippery Disease* (Washington, DC, 1978). Eminent influenza researcher Edwin Kilbourne termed influenza as an entity that is "at best transiently stable packages of genes borrowed from an extended gene pool," *Influenza*, p. 141.

TWO
The Murky Past of Influenza

1 The following terms and estimated dates are drawn from the Smithsonian Institution's Interactive Human Evolutionary Timeline. See http://humanorigins.si.edu, accessed 24 October 2011.

2 Based on the supposition that human ancestry extends to 5 million years ago and the Neolithic Revolution was in place 10,000 years ago.

3 The following discussion of the health implications of the transition from hunter-gatherer society to farming lifestyle is drawn from Mark Nathan Cohen, *Health and the Rise of Civilization* (New Haven, CT, 1989), William McNeill, *Plagues and Peoples* (New York, 1976), and Jared Diamond, *Guns, Germs, and Steel: The Fates of Human Societies* (New York, 1997).

4 All Indo-European languages share the root term *ghans* for goose, but do not have the same root name for ducks. The dispersal of the Indo-European language group is dated no later than 2500 BCE. The following discussion is drawn from Robert Langdon, "When the Blue-Egg Chickens Come Home to Roost: New Thoughts on the Prehistory of the Domestic Fowl in Asia, America, and the Pacific

Islands," *The Journal of Pacific History*, XXIV/2 (October 1989), pp. 164–92; James Harper, "The Tardy Domestication of the Duck," *Agricultural History*, XLVI/3 (July 1972), pp. 385–9; R. A. Donkin, *The Muscovy Duck, Carina Moschata Domestication: Origins, Dispersal and Associated Aspects of the Geography of Domestication* (Rotterdam, 1989); and E. Fuller Torrey and Robert H. Yolkey, *Beasts of the Earth: Animals, Humans and Disease* (New Brunswick, NJ, 2005) unless otherwise noted.

5 See Edward Hyams, *Animals in the Service of Man* (Philadelphia, 1972), pp. 33–4.

6 See Umberto Albarella, Keith Dobney, and Peter Rowley-Conwy, "The Domestication of the Pig (*Sus scrofa*): New Challenges and Approaches," in *Documenting Domestication: New Genetic and Archaeological Paradigms*, ed. Melinda A. Zeder, Daniel G. Bradley, Eve Emshwiller, and Bruce D. Smith (Berkeley, CA, 2006), pp. 209–27, for an overview of pig domestication; Sarah M. Nelson, ed., *Ancestors for the Pigs: Pigs in Prehistory* (Philadelphia, 1998); and J.-D. Vigne, J. Peters, and D. Helmer, eds, *First Steps of Animal Domestication: New Archeozoological Approaches*, Proceedings of the 9th ICAZ Conference, Durham, 2002 (Oxford, 2005) for an overview of animal domestication. See also Brett Mizelle, *Pig* (London, 2011) for an overview of pigs in history and culture.

7 For Chinese sites see Sarah M. Nelson, "Pigs in the Hongshan Culture," in *Ancestors for the Pigs*, pp. 99–107; for Korean and the Japanese islands see Akira Matsui, Naotaka Ishiguro, Hitomai Hongo, and Masao Minagawa, "Wild Pig? Or Domesticated Boar? An Archaeological View of the Domestication of *Sus scrofa* in Japan," in *First Steps of Animal Domestication*, ed. Vigne, Peters, and Helmer, pp. 148–59.

8 See Graeme Barker, *The Agricultural Revolution in Prehistory: Why did Foragers Become Farmers?* (Oxford, 2006), pp. 195–8.

9 The following discussion is drawn from McNeill's *Plagues and Peoples*, and *The Rise of the West: A History of the Human Community* (Chicago and London, 1963).

10 For Hippocrates see "The Cough of Perinthus," in Book VI: *Of the Epidemics in Hippocrates*, vol. VII, trans. and ed. Wesley D. Smith (Cambridge, MA, 1994), p. 269. For Livy see "The Siege of Achradina," in *The History of Rome in Livy*, 5 vols, trans. George Baker (New York, 1836), vol. III, pp. 96–8.

11 Pehr Osbeck in *A Voyage to China and the East Indies* [1771] as quoted by Donkin, *The Muscovy Duck*, p. 4.

12 For accounts identifying influenza in the Middle Ages see Warren T. Vaughan, *Influenza: An Epidemiologic Study* (Baltimore, MD, 1921), pp. 2–4, and W.I.B. Beveridge, "The Chronicle of Influenza Epidemics," *History and Philosophy of the Life Sciences*, XIII (1991), pp. 223–5. For Hirsch see August Hirsch, *Geographical and Historical Pathology*, vol. I: *Acute Infective Diseases*, trans. from the 2nd German edn Charles Creighton (London, 1883), pp. 7–54; chart is pp. 7–17.

13 An alternate explanation for the origin of the term stems from the seasonality of influenza outbreaks as being due to the "influence of the cold" (*influenza di freddo*). See W.I.B. Beveridge, *Influenza: The Last Great Plague; an Unfinished Story of Discovery* (New York, 1977), p. 24.

14 Theophilus Thompson, *Annals of Influenza or Epidemic Catarrhal Fever in Great Britain from 1510 to 1837* (London, 1852), quotes are from 3, 12, and 372. See also Charles Creighton, *A History of Epidemics in Britain*, vol. I: *From AD 664 to the Great Plague* [1894] (New York, 1965), pp. 568–77.

15 For a survey of Sydenham's account of the epidemic of 1675 see Charles Creighton, *A History of Epidemics in Britain*, vol. II: *From the Extinction of the Plague to the Present Time* [1891] (New York, 1965), pp. 326–9.

16 See John F. Townsend's discussion of the epidemic of 1781–2 which he drew from A. F. Hopkins, *Influenza: History, Nature, Courses and Treatments* [1913]. See John F. Townsend, "History of Influenza Epidemics," *Annals of Medical History*, V/6 (November 1933), pp. 537–8.

17 For the impact of William Farr on the field of epidemiology see John M. Eyler, *Victorian Social Medicine: The Ideas and Methods of William Farr* (Baltimore and London, 1979); Alexander Langmuir, "William Farr: Founder of Modern Concepts of Surveillance," *International Journal of Epidemiology*, V/1 (1976), pp. 13–18; and D. E. Lilienfeld, "Celebration: William Farr (1807–1883), An Appreciation on the 200th Anniversary of His Birth," *International Journal of Epidemiology*, XXXVI (2007), pp. 985–7.

18 The following discussion of the influenza epidemic of 1847–8 is drawn from Creighton, *A History of Epidemics in Britain*, vol. II, pp. 389–93.

THREE
Misidentifications and False Starts

1 See Alfred Crosby, *Ecological Imperialism: The Biological Expansion of Europe, 900–1900* (Cambridge, 1986), especially chapter Two, "Pangaea Revisited," pp. 8–40.

2 Jerry Bentley provides a brief overview of the formation of this system in Jerry H. Bentley, *Old World Encounters: Cross-cultural Contacts and Exchanges in Pre-Modern Times* (New York, 1993). For a comprehensive study see the six-volume work "History of Civilizations of Central Asia" sponsored by the United Nations Educational, Scientific, and Cultural Organization (UNESCO).

3 Unless otherwise noted, the following information about cholera is drawn from J. N. Hays, *The Burdens of Disease: Epidemics and Human Response in Western History* (New Brunswick, NJ, 2000), pp. 135–53; R. Pollitzer, *Cholera* (Geneva, 1959); R. S. Bray, *Armies of Pestilence: The Impact of Disease on History* (New York, 1996), pp. 154–92; and Charles Rosenberg, *The Cholera Years: The United States in 1832, 1849, and 1866* [1962] (Chicago, 1987).

4 Rosenberg, *The Cholera Years*, pp. 13–25.

5 The nineteenth-century strain known as "classic cholera" was especially virulent. The estimate of 50 per cent mortality is probably skewed high because there was little appreciation or identification of those who were mildly infected. Mortality estimate is from Hays, *Burdens of Disease*, p. 136. In the twentieth century the "classical" strain of cholera has been supplanted with a milder type known as "El Tor." El Tor occasions a less drastic infection which, combined with rehydration techniques and chemotherapies, has dropped cholera mortality to significantly lower levels. See J. Gallut, "The Cholera Vibrios," pp. 17–40, for the types of cholera and Norbert Hirschhorn, Nathaniel F. Pierce, Kazumine Kobari, and Charles C. J. Carpenter Jr, "The Treatment of Cholera," pp. 235–52, for medical techniques for those infected with cholera. Both essays are in *Cholera*, ed. Dhiman Barua and William Burrows (Philadelphia, 1974).

6 For the durability of the miasma concept see Carlo M. Cipolla, *Miasmas and Disease: Public Health and the Environment in the Pre-industrial Age*, trans. Elizabeth Potter (New Haven, CT, and London, 1992), especially

pp. 1–9; for an earlier environmentalist focus see James C. Riley, *The Eighteenth-century Campaign to Avoid Disease* (New York, 1987); for sanitarians, see John Duffy, *The Sanitarians: A History of American Public Health* (Urbana and Chicago, 1990). Sanitation practices had dramatic impact in improving health in the cities and cutting the incidence of certain diseases, especially enteric ones like typhoid. The sanitarians were "right" about how to improve health but for the wrong reasons.

7 Von Pettenkofer and his disciples clung to this explanation into the twentieth century. To demonstrate the surety of his explanation, von Pettenkofer proffered a dramatic display. He drank a pure culture of the purported cholera vibrio, suffering only mild complaints. His assistant did not fare as well in duplicating this bold exhibition as he endured debilitating diarrheal symptoms. For the XYZ explanation see Hays, *Burdens of Disease*, pp. 148–9; for von Pettenkofer's dramatic experiment see ibid., pp. 151–2, and C.E.A. Winslow, *Man and Epidemics* (Princeton, NJ, 1952), pp. 38–9.

8 There may be some validity to the hypothesis that the poor are more susceptible to cholera infection beyond a greater likelihood of exposure to the vibrio. Malnutrition is associated with decreased acid production in the stomach. Such a situation may make it more probable that cholera could survive the destructive trip through the stomach and thus increase the rate of infection among the poor. Such a hypothesis cannot be humanely tested however and unraveling this phenomenon from the background of a cholera outbreak is devilishly difficult. See Eugene J. Gangarosa and Wiley H. Mosley, "Epidemiology and Surveillance of Cholera," in *Cholera*, ed. Barua and Burrows, pp. 381–403.

9 See Alan Kraut, *Silent Travelers: Germs, Genes, and the "Immigrant Menace"* (New York, 1994), and Howard Markel, *When Germs Travel: Six Major Epidemics that have Invaded America since 1900 and the Fears they have Unleashed* (New York, 2004).

10 See Hays, *Burdens of Disease*, pp. 141–53; George Rosen, *A History of Public Health* (New York, 1958), pp. 294–343; Duffy, *The Sanitarians*, pp. 193–220; and Winslow, *Man and Epidemics*, pp. 17–23.

11 For Snow see Rosenberg, *The Cholera Years*, pp. 193–4, and Steven Johnson, *The Ghost Map: The Story of London's Most Terrifying Epidemic and How it Changed Science, Cities, and the Modern World* (New York, 2006).

12 The following discussion of the early appearance of the Russian flu of
    1889 comes from H. Franklin Parsons, *Report on the Influenza Epidemic of
    1889–90: Presented to Both Houses of Parliament by Command of Her Majesty*
    (London, 1891), pp. 1–14. It should be noted that K. David Patterson
    raises some questions about the accuracy of the diagnosis of influenza
    at Bukhara. See K. David Patterson, *Pandemic Influenza, 1700–1900:
    A Study in Historical Epidemiology* (Totowa, NJ, 1986), p. 52.

13 Chronology of the flu from Parsons, *Report*, p. 71.

14 See Mari Loreena Nicholson-Preuss, "Managing Morbidity and
    Mortality: Pandemic Influenza in France, 1889–90," unpublished MA
    Thesis in History, Texas Tech University (2001).

15 For France, ibid., p. 62. For Barcelona see Esteban Rodriguez-Ocana,
    "Barcelona's Influenza: A Comparison of the 1889–90 and 1918
    Autumn Outbreaks," in *Influenza and Public Health: Learning from Past
    Pandemics*, ed. Tamara Giles-Vernick and Susan Craddock, with
    Jennifer Gunn (London, 2010), p. 57.

16 A quick scan of newspaper databases from early 1890 will turn up
    any number of liquids, powders, and plasters guaranteed to gird the
    consumer against influenza. A personal favorite is the mysterious
    and magical "electricity in the bottle." See advertisement for "West's
    Electric Cure" collected from *The Daily Inter Ocean* [Chicago] (29
    December 1889), p. 11. Collected from *America's Historical Newspaper
    Database, Newsbank*, accessed 25 August 2008.

17 See Patterson for an assessment of the variety of sources. Patterson,
    Pandemic Influenza, pp. 49–82.

18 See Mark Honigsbaum, "The 'Russian' Influenza in the UK: Lessons
    Learned, Opportunities Missed," *Vaccine*, 29S (2011), pp. B11–B15.

19 Patterson, *Pandemic Influenza*, p. 72. The lack of data from regions
    outside of Europe prevented Patterson from extrapolating the pattern
    to other regions of the world in order to generate a global total. Using
    the figures from the Medical Department of the Local Government
    Board, Honigsbaum reports that no fewer than 125,000 died in
    England and Wales during the first and subsequent waves of the
    Russian flu (1890–1893). Honigsbaum, "The 'Russian' Influenza in
    the UK," p. B12.

20 Pfeifer's bacillus is now known as *Haemophilus influenzae* or alternately
    Bacillus influenzae. For Pfeiffer's biographical information see Paul

Fildes, "Richard Friedrich Johannes Pfeiffer, 1858–1945," *Biographical Memoirs of the Royal Society*, 11 (November 1956), pp. 237–47; for acceptance of Pfeiffer's theory of influenza causation see R. Thorne, "Introduction by the Medical Officer," in Local Government Board, *Further Report and Papers on Pandemic Influenza, 1889–92: Presented to Both Houses of Parliament by Command of Her Majesty* (London, 1893), pp. vii–x.

21 For a description of this process in an American context see Nancy Tomes, *The Gospel of Germs: Men, Women, and the Microbe in American Life* (Cambridge, MA, 1998).

22 "Spanish flu" has generated an amazing quantity of books and articles in the last decade of the twentieth and the first of the twenty-first centuries. Unless otherwise noted, the following discussion of Spanish flu is drawn from Edwin Oakes Jordan, *Epidemic Influenza: A Survey* (Chicago, 1927); Alfred Crosby, *Epidemic and Peace, 1918* (Westport, CT, 1976). Crosby's book was subsequently reissued and can also be found under the title *America's Forgotten Pandemic* (Cambridge, 1989); Howard Phillips and David Killingray, eds, *The Spanish Influenza Pandemic of 1918–19: New Perspectives* (London, 2003); John M. Barry, *The Great Influenza: The Epic Story of the Deadliest Plague in History* (New York, 2004); and Carol R. Byerly, *Fever of War: The Influenza Epidemic in the U.S. Army During World War I* (New York, 2005).

23 An alternative origin for the affliction, vigorously promoted by virologist J. S. Oxford, maintains that the first cases of infection emerged at military hospitals near the Western Front during the First World War. Physicians there diagnosed an unusual illness they called "purulent bronchitis." See J. S. Oxford, A. Sefton, R. Jackson, N.P.A.S Johnson, and R. S. Daniels, "Who's That Lady?," *Nature Medicine*, V/12 (December 1999), pp. 1351–2; J. S. Oxford, "The So-called Great Spanish Influenza Pandemic of 1918 may have Originated in France in 1916," *Philosophical Transactions of the Royal Society of London, Series B*, CCCLVI (2001), pp. 1857–9; and J. S. Oxford, R. Lambkin, A. Sefton, R. Daniels, A. Elliot, R. Brown, and D. Gill, "A Hypothesis: the Conjunction of Soldiers, Gas, Pigs, Ducks, Geese, and Horses in Northern France During the Great War Provided the Conditions for the Emergence of the 'Spanish' Influenza Pandemic of 1918–19," *Vaccine*, XXIII/7 (January 2005), pp. 940–45.

24 Carol Byerly's book *Fever of War* discusses the U.S. medical efforts in the Army during the war.

25 For account of Sheffield see Mark Honigsbaum, "The Great Dread: Cultural and Psychological Impacts and Responses to the 'Russian' Influenza in the United Kingdom, 1889–1893," *Social History of Medicine*, XXIII/2 (August 2010), pp. 299–319.

26 For example, 790,000 of the 2 million U.S. doughboys disembarked at the port. Crosby, *Epidemic and Peace*, p. 38.

27 The name "Spanish" was applied to the pandemic because the first wave of the infection was widely reported in the Spanish press. Spain, neutral in the war, had no press censorship and thus was able to discuss openly the infection circulating in the state which attacked a number of people including the King. See Crosby, *Epidemic and Peace*, p. 26, and Barry, *The Great Influenza*, p. 171.

28 See Jeffrey K. Taubenberger, Ann H. Reid, Thomas A. Janczewski, Thomas G. Fanning, "Integrating Historical, Clinical, and Molecular Genetic Data in Order to Explain the Origin and Virulence of the 1918 Spanish Influenza Virus," *Philosophical Transactions: Biological Sciences*, CCCLVI/1416 (29 December 2001), pp. 1829–39, who posit an estimated 28 per cent global infection rate. Other assays range even higher.

29 Unless otherwise noted, the following discussion of Camp Devens is drawn from Crosby, *Epidemic and Peace*, pp. 3–11, and Jordan, *Epidemic Influenza*, pp. 97–118. See especially table 22, on p. 101.

30 Figures for pneumonia cases at Camp Devens are drawn from Jordan, *Epidemic Influenza* (table 28), p. 116. Pathological examination of the lungs from selected fatal cases revealed severe damage to the bronchi and alveoli. See S. Burt Wolbach and Channing Frothingham, "The Influenza Epidemic at Camp Devens in 1918: A Study of the Pathology of the Fatal Cases," *Archives of Internal Medicine*, XXXII/4 (October 1923), pp. 571–600.

31 Pneumonia and influenza mortality from Jordan, *Epidemic Influenza*, p. 101. Epidemiologists refer to these mortality numbers as the P&I figures.

32 Editorial, *American Journal of Public Health*, VIII (October 1918), pp. 787–8. Citation suggested from Crosby, *Epidemic and Peace*, p. 92.

33 My office at my home university (Wichita State) was a dormitory that was repurposed as a temporary Spanish flu hospital at the (then

Fairmount) College and for the surrounding neighborhood.

34 Australia managed to delay the pandemic's arrival through quarantine efforts, but eventually the affliction appeared, likely brought by returning First World War personnel. Only American Samoa and a few isolated communities who zealously maintained their isolation managed to escape the Spanish Flu's ravages. See Crosby, *Epidemic and Peace*, pp. 234–41.

35 John Barry's *The Great Influenza* details this desperate research effort in the United States. See Barry, *The Great Influenza*, generally, but especially 253–94.

36 N. Montefusco in *Riforma Medica*, Naples, XXXIV/28 (13 July 1918), p. 549, as quoted in Current Medical Literature, JAMA, LXXI/11 (14 September 1918), p. 934. Citation suggested by Barry, *The Great Influenza*, p. 265. Montefusco was obviously discussing the spring wave of Spanish flu in this correspondence.

37 Gary Gernhart, Office of PHS Historian, "A Forgotten Enemy: PHS's Fight Against the 1918 Influenza Pandemic," *Public Health Reports*, CXIV (December 1999), pp. 559–61.

38 For example, much of the research on influenza at the Rockefeller Institute still revolved around *Bacillus influenza* in the immediate aftermath of the pandemic. See Saul Benison's oral history memoir, *Tom Rivers: Reflections on a Life in Medicine and Science* (Cambridge, MA, 1967), pp. 67–150. The oral history is a fascinating account of research and administration from one of the founding fathers of virology.

39 See Michael Charles Bresalier, "Transforming Flu: Medical Science and the Making of Virus Disease in London, 1890–1939," PhD dissertation, Trinity College, Cambridge (2010), pp. 176–200.

40 Crosby's famous "forgetting" formulation can be found in Crosby, *Epidemic and Peace*, pp. 311–28.

41 See Byerly, *Fever of War*, generally on this issue, but especially pp. 14–38 and 125–52.

42 For an overview of this discussion see Howard Phillips and David Killingray, "Introduction," in *The Spanish Influenza Pandemic*, ed. Phillips and Killingray, pp. 1–25.

43 Jordan's *Epidemic Influenza* provides the richest collection of citations; Crosby provides an accessible discussion of the general mortality pattern in *Epidemic and Peace*, pp. 203–26.

44 For example, South African cities reported that 60 per cent of their deaths were in the decades encapsulating 20–40 years of age. See Barry, *The Great Influenza*, p. 239. In Sydney people in the age bracket of the mid-twenties and mid-thirties suffered the highest mortality rate in that city. See Kevin McCracken and Peter Curson, "Flu Downunder: A Demographic and Geographic Analysis of the 1919 Epidemic in Sydney, Australia," in *The Spanish Influenza Pandemic*, ed. Phillips and Killingray, pp. 110–31. The same pattern can be observed in records from Britain. See N.P.A.S. Johnson, "The Overshadowed Killer: Influenza in Britain in 1918–19," in *The Spanish Influenza Pandemic*, ed. Phillips and Killingray, pp. 132–55.

45 See Jordan, *Epidemic Influenza*, for his prodigious effort at evaluating the Spanish flu's impact.

46 See Appendix B from Kingsley Davis, *The Population of India and Pakistan* (Princeton, NJ, 1951), p. 237.

47 The following estimate of Spanish flu mortality is drawn from K. David Patterson and Gerald F. Pyle, "The Geography and Mortality of the 1918 Influenza Pandemic," *Bulletin of the History of Medicine*, LXV (1991), pp. 4–21.

48 See Niall P.A.S. Johnson and Juergen Mueller, "Updating the Accounts: Global Mortality of the 1918–1920 'Spanish' Influenza Pandemic," *Bulletin of the History of Medicine*, LXXVI/1 (2002), pp. 105–15.

49 After his mammoth survey of the medical literature on the pandemic, Edwin Oakes Jordan mournfully concluded that "In the face of the almost certain recurrence some day of another world-wide pandemic, we remain nearly as helpless to institute effective measures of control as we were before 1918." Jordan, *Epidemic Influenza*, p. 3.

50 See W.I.B. Beveridge, *Influenza: The Last Great Plague; An Unfinished Story of Discovery* (New York, 1977).

FOUR
The 1920s to the 1980s

1 See British pathologist Robert Donaldson's "The Bacteriology of Influenza: With Special Reference to Pfeiffer's Bacillus," in *Influenza: Essays by Several Authors*, ed. F. G. Crookshank (London, 1922), pp. 139–236.

2 See Dorothy H. Crawford, *The Invisible Enemy: A Natural History of Viruses* (Oxford, 2000), pp. 12–13.

3 Michael Charles Bresalier, "Transforming Flu: Medical Science and the Making of a Virus Disease in London, 1890–1939," PhD dissertation, Trinity College, Cambridge (2010), p. 143.

4 For an overview of these experiments up to the 1920s see Donaldson, "The Bacteriology of Influenza," pp. 230–36.

5 Bresalier, "Transforming Flu," p. 151.

6 See W.I.B. Beveridge, *Influenza: The Last Great Plague; An Unfinished Story of Discovery* (New York, 1977), pp. 4–5, and David Tyrrell, "Discovery of Influenza Viruses," in *Textbook of Influenza*, ed. Karl G. Nicholson, Robert G. Webster, and Alan J. Hay (London, 1998), p. 21.

7 Richard Shope, "Swine Influenza I: Experimental Transmission and Pathology," *Journal of Experimental Medicine*, LIV (1931), pp. 349–60; Richard Shope, "Swine Influenza II: A Hemophilic Bacillus from the Respiratory Tract of Infected Swine," *Journal of Experimental Medicine*, LIV (1931), pp. 361–72; and Richard Shope, "Swine Influenza III: Filtration Experiments and Etiology," *Journal of Experimental Medicine*, LIV (1931), pp. 373–85.

8 For the growth of the MRC see Bresalier, "Transforming Flu," pp. 101–10.

9 See Michael Bresalier, "Neutralizing Flu: 'Immunological Devices' and the Making of a Virus Disease," in *Crafting Immunity: Working Histories of Clinical Immunology*, ed. Kenton Kroker, Pauline M. H. Mazumdar, and Jennifer Keelan (Aldershot, 2008), pp. 114–17.

10 The following account of the demonstration that influenza is a viral disease by the use of ferrets as an animal model is drawn from Beveridge, *Influenza*, pp. 7–9, and Bresalier, "Neutralizing Flu," pp. 121–9.

11 Stuart-Harris's exposure in 1936 was potentially the second laboratory animal-to-human transference of the illness. In 1934, Thomas Francis, an American influenza researcher, reported that "In the course of the experimental work with ferrets, one of the laboratory workers (S. S.) developed symptoms typical of influenza." Thomas Francis, "Transmission of Influenza by a Filterable Virus," *Science*, n.s. LXXX/2081 (16 November 1934), p. 458. For Stuart-Harris's infection see Bresalier, "Transforming Flu," p. 256.

12 See Beveridge, *Influenza*, pp. 8–9, and Tyrrell, "The Discovery of Influenza Viruses," pp. 23–4.

13 In fact, Andrewes and Shope developed a correspondence friendship that lasted for decades. See Tyrrell, "The Discovery of Influenza Viruses," pp. 22–3.

14 The first vaccine tests were conducted on a small scale on institutionalized children located at state homes in the United States. These pilot tests were conducted by Joseph Stokes, a pediatrician at the University of Pennsylvania. See John M. Eyler, "De Kruif's Boast: Vaccine Trials and the Construction of a Virus," *Bulletin of the History of Medicine*, LXXX/3 (Fall 2006), pp. 410, 417–20, and Thomas Francis Jr, "Vaccination Against Influenza," *Bulletin of the World Health Organization*, VIII (1953), pp. 725–9. Discussion of the British vaccine trials comes from Bresalier, "Transforming Flu," pp. 258–63.

15 Measuring antibodies relied upon collecting two serum samples from inoculated mice to compare before and after reaction rates. The process was time-consuming as the mice were infected, serum was drawn and then the mice were re-exposed at least two weeks later. The process was also labor-intensive as the lungs of each mouse would have to be examined individually to gauge the response rate. Typically the test would require about three weeks to generate results. Finally, a legion of mice was needed using different volumes of material in order to effectively measure antibody response. For example, a U.S. test of a vaccine, measuring the response rate of 248 vaccinated people and 60 people stricken naturally required the use of 11,000 mice. Clearly this was not an efficient or cost-effective method of detecting antibody protection formulas. See Eyler, "De Kruif's Boast," p. 416.

16 Ibid., pp. 421–3.

17 See John R. Paul, *A History of Poliomyelitis* (New Haven, CT, and London, 1971), pp. 413–17.

18 See Tyrrell, "The Discovery of Influenza Viruses," pp. 23–6, and Eyler, "De Kruif's Boast," p. 417.

19 For concentrating vaccines and measuring reaction rates see Thomas A. Francis and Jonas E. Salk, "A Simplified Procedure for the Concentration and Purification of Influenza Virus," *Science*, n.s. XCVI/2500 (27 November 1942), pp. 499–500, and Jonas E. Salk,

"Reactions to Concentrated Influenza Virus Vaccines," *Journal of Immunology*, 58 (1948), pp. 369–95.

20 The report of the vaccine trial of 1943 can be found at Members of the Commission on Influenza, Board for the Investigation and Control of Influenza and other Epidemic Diseases in the Army, Preventive Medical Service, Office of the Surgeon General, United States Army, "A Clinical Evaluation of Vaccination Against Influenza: Preliminary Report," *Journal of the American Medical Association*, CXXIV/14 (1 April 1944), pp. 982–5. See also John M. Wood and Michael S. Williams, "History of Inactivated Influenza Vaccines," in *Textbook of Influenza*, ed. Nicholson, Webster, and Hay, pp. 317–23.

21 The shorthand name for the strains was initially a mixture of either the place where the virus was recovered (PR8: Puerto Rico) or for the person it was recovered from (WS: Wilson Smith). Later these names became standardized with the type, location, year, and if necessary, the species from which the sample was collected. For example, Swine flu (1976) was labeled A/New Jersey/76.

22 Thomas Francis Jr, Jonas E. Salk, and J. J. Quilligan Jr, "Experience with Vaccination Against Influenza in the Spring of 1947: A Preliminary Report," *American Journal of Public Health*, XXXVII (August 1947), pp. 1013–16.

23 The changed virus in 1947 was later called an intrasubtypic antigenic change. See Edwin Kilbourne, Catherine Smith, Ian Brett, Barbara A. Pokorny, Bert Johansson, and Nancy Cox, "The Total Influenza Vaccine Failure of 1947 Revisited: Major Intrasubtypic Antigenic Change can Explain Failure of Vaccine in a Post-World War II Epidemic," *Proceedings of the National Academy of Sciences*, XCIX/16 (6 August 2002), pp. 10748–52.

24 For discussion of the 1946 Influenza B and the 1947 Influenza A epidemics and the founding of the World Influenza Centre see C. H. Andrewes, "Epidemiology of Influenza," *Bulletin of the World Health Organization*, VIII (1953), pp. 600–05, and A.M.-M. Payne, "The Influenza Programme of WHO," *Bulletin of the World Health Organization*, VIII (1953), pp. 756–60.

25 See James T. Culbertson, "Plans for United States Cooperation with the World Health Organization in the International Influenza Study Program," *American Journal of Public Health*, XXXIX (January 1949),

pp. 37–43, and Dorland J. Davis, "World Health Organization Influenza Study Program in the United States," *Public Health Reports*, LXVII/12 (December 1952), pp. 1185–90.

26 Payne, "The Influenza Programme of WHO," p. 760.

27 Ian G. S. Furminger, "Vaccine Production," in *Textbook of Influenza*, ed. Nicholson, Webster, and Hay, pp. 324–6.

28 Maurice Hilleman, "Six Decades of Vaccine Development: A Personal History," *Nature Medicine Vaccine Supplement*, IV/5 (May 1998), pp. 507–14; Paul A. Offit, *Vaccinated: One Man's Quest to Defeat the World's Deadliest Diseases* (New York, 2007), pp. 1–19; and "Hong Kong Battling Influenza Epidemic," *New York Times* (17 April 1957), p. 3.

29 Offit, *Vaccinated*, p. 13.

30 Continuing the tradition of naming new influenza strains after the regions in which they are first discovered, the new virus was initially dubbed Far Eastern influenza. This name was subsequently amended to the more accurate Asian flu.

31 The Communicable Disease Center later became the Center for Disease Control and still later the Centers for Disease Control and Prevention; the name it operates under today. The initials "CDC" have remained unchanged throughout its tenure.

32 The mistaken man-made infectious outbreak is referred to as the "Cutter Incident" in dubious honor of the pharmaceutical firm responsible for the majority of the cases. See Paul Offit, *The Cutter Incident: How America's First Polio Vaccine Led to the Growing Vaccine Crisis* (New Haven, CT, and London, 2005), and Mark Pendergrast, *Inside the Outbreaks: The Elite Medical Detectives of the Epidemic Intelligence Service* (Boston, 2010), pp. 21–6. For Langmuir's depiction of the EIS as a medical fire department see D. A. Henderson, *Smallpox: The Death of a Disease* (Amherst, NY, 2009), pp. 23–4.

33 These reports were issued as frequently as every three days and proved very useful for alerting public health experts to the spread of the pandemic. See D. A. Henderson, Brooke Courtney, Thomas V. Inglesby, Eric Toner, and Jennifer B. Nuzzo, "Public Health and Medical Response to the 1957–58 Influenza Pandemic," *Biosecurity and Bioterrorism: Biodefense Strategy, Practice, and Science*, VII/3 (2009), pp. 265–73. These reports have been collected and bound at the National Library of Medicine at Bethesda, Maryland.

34 Offit, *Vaccinated*, pp. 14–15.

35 See Elizabeth W. Etheridge, *Sentinel for Health: A History of the Centers for Disease Control* (Berkeley, CA, 1992), pp. 80–86.

36 J. Mulder, "Asiatic Influenza in the Netherlands," *Lancet*, CCLXX/6990 (17 August 1957), p. 334.

37 F. M. Davenport et al., "Further Observations of the Relevance of Serologic Recapitulations of Human Infection with Influenza Viruses," *Journal of Experimental Medicine*, CXX (1964), pp. 1087–97.

38 These national summaries are stored as microfiche at the World Health Organization's Archives, Geneva, filed 12/418/12 (hereafter WHO Archives, Geneva).

39 See 21 May 1957 letter from Dr C. Mani, Regional Director, SEARO, to Dr W. Timmerman, ADC-CTS 12/418/12 2 (microfiche), WHO Archives, Geneva.

40 U.S. health officials estimated that 25 per cent of the population was infected with Asian flu just in the months of October and November. See Henderson et al., "Public Health and Medical Response to the 1957–58 Influenza Pandemic," p. 271. For global mortality see Derek J. Smith, "Predictability and Preparedness in Influenza Control," *Science*, CCCXII (21 April 2006), p. 392.

41 The following information about the discovery of Hong Kong flu is drawn from W. Charles Cockburn, P. J. Delon, and W. Ferreira, "Origin and Progress of the 1968–69 Hong Kong Influenza Epidemic," *Bulletin of the World Health Organization*, XLI (1969), pp. 345–8. It should be noted that a search of the *Times* (London) archives did not turn up an article reporting on an influenza outbreak in China on that date or on any date close to 12 July 1968. Perhaps the detail of the newspaper he was reading escaped Cockburn.

42 The exchange of telegrams between Chang and Cockburn can be found at Folio, "Information on Influenza Incidences (1 July 1968–30 June 1969)," 12/442/2 (68–69), Jacket No. 1, WHO Archives, Geneva.

43 See Cecile Viboud, Rebecca F. Grais, Bernard A. P. Lafont, Mark A. Miller, and Lone Simonsen, "Multinational Impact of the 1968 Hong Kong Influenza Pandemic: Evidence for a Smoldering Pandemic," *Journal of Infectious Diseases*, CXCII (15 July 2005), pp. 233–48.

44 The conference papers are collected in the *Bulletin of the World Health Organization*, XLI/3–4–5 (1969), pp. 345–748.

45 See for example F. M. Davenport, E. Minuse, A. V. Hennessy, and
    T. Francis Jr, "Interpretations of Influenza Antibody Patterns of Man,"
    Bulletin of the World Health Organization, XLI (1969), pp. 453–60;
    N. Masurel, "Serological Characteristics of a 'New' Serotype of
    Influenza A Virus: The Hong Kong Strain," *Bulletin of the World Health
    Organization*, XLI (1969), pp. 461–8; Hideo Fukumi, "Interpretation
    and Significance of Hong Kong Antibody in Old People Prior to the
    Hong Kong Influenza Epidemic," *Bulletin of the World Health
    Organization*, XLI (1969), pp. 469–73.

46 See Edwin Kilbourne, "Future Influenza Vaccines and the Use of
    Genetic Recombinants," *Bulletin of the World Health Organization*, XLI
    (1969), pp. 643–5.

47 The naming system remains the one utilized today. For a full
    discussion of why the new naming system was needed see Walter
    R. Dowdle, Marion T. Coleman, Elmer C. Hall, and Violeta Knez,
    "Properties of the Hong Kong Influenza Virus," *Bulletin of the World
    Health Organization*, XLI (1969), pp. 419–24.

48 Mortality totals from Smith, "Predictability and Preparedness in
    Influenza Control," p. 392.

49 Two standard works on the Swine flu affair are Richard E. Neustadt
    and Harvey V. Fineberg, *The Swine Flu Affair: Decision-making on a Slippery
    Disease* (Washington, DC, 1978), and Arthur M. Silverstein, *Pure Politics
    and Impure Science: The Swine Flu Affair* (Baltimore and London, 1981).

50 In 2004, the U.S. National Academy of Sciences took up the dispute
    and concluded that the evidence "favored acceptance of a causal
    relationship" between the Swine flu vaccine and increased rates
    of GBS. See Kathleen Stratton, Donna A. Alamario, Theresa
    Wizemann, and Marie C. McCormick, eds, *Immunization Safety
    Review: Influenza Vaccines and Neurological Complications* (Washington,
    DC, 2004), p. 1.

51 See Alan P. Kendal, Gary R. Noble, John J. Skehel, and Walter
    R. Dowdle, "Antigenic Similarity of Influenza A (H1N1) Virus from
    Epidemics in 1977–1978 to 'Scandinavian' Strains Isolated in
    Epidemics of 1950–1951," *Virology*, LXXXIX (1978), pp. 632–6.

FIVE
Renewed Fears of Flu

1 Macfarlane Burnet and David O. White, *Natural History of Infectious Disease*, 4th edn, as quoted in Gerald N. Grob, *The Deadly Truth: A History of Disease in America* (Cambridge, MA, 2002), pp. 272–3.

2 The conference prompted an Institute of Medicine Report, *Emerging Infections: Microbial Threats to Health in the United States* (Washington, DC, 1992) and an edited collection from one of its strongest boosters, Stephen S. Morse, ed., *Emerging Viruses* (New York, 1993).

3 The following information about Taubenberger, the Institute of Pathology and the recovery of the Spanish flu is drawn from Patricia Gadsby, "Fear of Flu: Pandemic Influenza Outbreaks," *Discover*, XX/I (January 1999), p. 82; Diane Martindale, "No Mercy," *New Scientist* (14 October 2000), p. 2929; Gina Kolata, *Flu: The Story of the Great Influenza Pandemic of 1918 and the Search for the Virus that Caused It* (New York, 1999), pp. 187–218; Pete Davies, *The Devil's Flu: The World's Deadliest Influenza Epidemic and the Scientific Hunt for the Virus that Caused It* (New York, 2000), pp. 194–223; and Jeffrey K. Taubenberger, Ann H. Reid, Amy E. Krafft, Karen E. Bijwaard, and Thomas G. Fanning, "Initial Genetic Characterization of the 1918 'Spanish' Influenza Virus," *Science*, CCLXXV (21 March 1997), pp. 1793–6.

4 For Hultin's story see Alfred Crosby, *Epidemic and Peace, 1918* (Westport, CT, 1976), pp. 305–6; Kolata, *Flu*, pp. 85–120; and Davies, *The Devil's Flu*, pp. 224–49.

5 The description of the outbreak on the farms in early 1997 is drawn from L. D. Sims, T. M. Ellis, K. K. Liu, K. Dyrtirg, H. Wong, M. Peiris, Y. Guan, and K. F. Shortridge, "Session Keynote Address: Avian Influenza in Hong Kong, 1997–2002," *Avian Diseases*, XLVII, Special Issue, Proceedings of the Fifth International Symposium on Avian Influenza (2003), pp. 832–8; email exchange with Leslie Sims, 17 June 2011; and Kennedy Shortridge, personal interview, 17 June 2011.

6 Survey of HPAI outbreaks from Michael L. Perdue and David E. Swayne, "Invited Minireview: Public Health Risk from Avian Influenza Viruses," *Avian Diseases*, XLIX/3 (September 2009), pp. 317–27.

7 Leslie D. Sims and Andrew J. Turner, "Avian Influenza in Australia," in *Avian Influenza*, ed. David Swayne (Ames, IA, 2008), pp. 239–50.

8 For an overview of this research see William Graeme Laver, Norbert Bischofberger, and Robert G. Webster, "The Origin and Control of Pandemic Influenza," *Perspectives in Biology and Medicine*, XLIII/2 (Winter 2000), pp. 173–92, and Robert G. Webster and William J. Bean Jr, "Evolution and Ecology of Influenza Viruses: Interspecies Transmission," in *Textbook of Influenza*, ed. Karl G. Nicholson, Robert G. Webster, and Alan J. Hay (London, 1998), pp. 109–19.

9 The following information on events in Hong Kong through August 1997 comes from J. C. de Jong, E.C.J. Claas, A.D.M.E. Osterhaus, R. G. Webster, and W. I. Lim, "A Pandemic Warning?," *Nature*, CCCLXXXIX (9 October 1997), p. 554; Kanta Subbarao et al., "Characterization of an Avian Influenza A (H5N1) Virus Isolated from a Child with a Fatal Respiratory Illness," *Science*, CCLXXIX /5349 (16 January 1998), pp. 393–6; Davies, *The Devil's Flu*, pp. 7–20; Michael Greger, *Bird Flu: A Virus of Our Own Hatching* (New York, 2006), pp. 32–5; Kennedy Shortridge, personal interviews, 17 June 2011 and 19 June 2011; and A.S.W. Ku and L.T.W. Chan, "The First Case of H5N1 Avian Influenza Infection in a Human with Complications of Adult Respiratory Distress Syndrome and Reye's Syndrome," *Journal of Paediatric Child Health*, XXXV (1999), pp. 207–9.

10 De Jong et al., "A Pandemic Warning?", p. 554.

11 Kennedy Shortridge, personal interview, 19 June 2011.

12 Unless otherwise noted, the following description of events in Hong Kong during November and December 1997 is drawn from Davies, *The Devil's Flu*, pp. 20–38; K. F. Shortridge, J.S.M. Peiris, and Y. Guan, "The Next Influenza Pandemic: Lessons from Hong Kong," *Journal of Applied Microbiology*, XCIV (2003), pp. 70S–79S; Anthony W. Mounts et al., "Case-Control Study of Risk Factors for Avian Influenza A (H5N1) Disease, Hong Kong, 1997," *The Journal of Infectious Diseases*, CLXXX (1999), pp. 505–8; Sims et al., "Avian Influenza in Hong Kong," p. 832–8; and Kennedy Shortridge, personal interviews, 17 June 2011 and 19 June 2011.

13 Leslie Sims, email exchange with author, 17 June 2011.

14 As Kennedy Shortridge stated, it's a "very old style life. I used to enjoy it. You get the feeling of the closeness of humans and animals and I like it very much . . . but you could almost smell the origins of a pandemic." Kennedy Shortridge, personal interview, 17 June 2011.

15 Leslie Sims, email exchange with author, 17 June 2011.

16 Kennedy Shortridge, personal interview, 19 June 2011.

17 Robert Webster, personal interview, 2 June 2011, and Davies, *The Devil's Flu*, pp. 17–38.

18 Kennedy Shortridge, personal interview, 19 June 2011.

19 Kennedy Shortridge, Peng Gao, Yi Guan, Toshihiro Ito, Yoshihiro Kawaoka, Deborah Markwell, Ayato Takada, and Robert G. Webster, "Interspecies Transmission of Influenza Viruses: H5N1 Virus and a Hong Kong SAR Perspective," *Veterinary Microbiology*, LXXIV (2000), p. 143.

20 Ibid.

21 Leslie D. Sims and Ian H. Brown, "Multicontinental Epidemic of H5N1 HPAI Virus (1996–2007)," in *Avian Influenza*, ed. David Swayne (Ames, IA, 2008), p. 271.

22 Kennedy Shortridge, personal interview, 19 June 2011.

23 See Paul K. S. Chan, "Outbreak of Avian Influenza A (H5N1) Virus Infection in Hong Kong in 1997," *Clinical Infectious Diseases*, XXXIV, Suppl. 2 (2002), pp. s58–s64, and Angela N. Cauthen, David E. Swayne, Stacey Schultz-Cherry, Michael L. Perdue, and David L. Suarez, "Continued Circulation in China of High Pathogenic Avian Influenza Viruses Encoding the Hemagglutinin Gene Associated with the 1997 H5N1 outbreak in Poultry and Humans," *Journal of Virology*, LXXIV/14 (July 2000), pp. 6592–9.

24 100–200 million figure in Perdue and Swayne, "Invited Minireview: Public Health Risk from Avian Influenza Viruses," see especially table on p. 320.

25 Unless otherwise noted, this and the following information about Bird flu is drawn from J. S. Malik Peiris, Menno D. de Jong, and Yi Guan, "Avian Influenza Virus (H5N1): A Threat to Human Health," *Clinical Microbiology Reviews*, XX/2 (2007), pp. 243–67.

26 Kennedy Shortridge, personal interview, 19 June 2011.

27 On 14 December 1949, Australian veterinarian W.I.B. Beveridge sent a letter to WIC Director at Mill Hill, C. H. Andrewes, reporting the text of a letter sent from James H. Steele, Chief, Veterinary Public Health Division of the United States Public Health Service. Steele said that the U.S. had been avidly studying swine influenza strains to determine "if they are of value in developing human vaccine." The U.S. would of

course be willing to send samples to London. See letter from W.I.B. Beveridge to C. H. Andrewes dated 14 December 1949, WHO Archives, WHO 2, D. C. INFL. 6, WIC (microfiche).

28 See International Health Regulations, available at www.who.int/en, accessed 28 June 2011.

29 L. D. Sims, "Lessons Learned from Asian H5N1 Outbreak Control," *Avian Diseases*, LI/1, Supplement: Sixth International Symposium on Avian Influenza (March 2007), pp. 174–81.

30 Estimates of SARS deaths up through 31 July 2003, "Summary of Probable SARS Cases with Onset of Illness from 1 November 2002 to 31 July 2003," available at www.who.int/en, accessed 1 June 2012. For the estimated costs of SARS see Richard D. Smith, "Responding to Global Infectious Disease Outbreaks: Lessons from SARS on the Role of Risk Perception, Communication, and Management," *Social Science and Medicine*, LXIII/12 (2006), p. 3114.

31 Robert F. Breiman, Meirion R. Evans, Wolfgang Preiser, James Maguire, Alan Schnur, Ailan Li, Henk Bekedam, and John S. MacKenzie, "Role of China in the Quest To Define and Control Severe Acute Respiratory Syndrome," *Emerging Infectious Diseases*, IX/9 (September 2003), and David L. Heymann and Guenael Rodier, "Global Surveillance, National Surveillance, and SARS," *Emerging Infectious Diseases*, X/2 (February 2004), accessed 31 October 2004.

SIX
Yet Another Surprise

1 The following information about the discovery of Swine flu (2009) is from the CDC reports in the *Morbidity and Mortality Weekly Report*, "Swine Influenza (H1N1) Infection in Two Children—Southern California, March–April 2009," *MMWR*, LVIII/15 (24 April 2009), pp. 400–02; "Update: Swine Influenza A (H1N1) Infections—California and Texas, April 2009," *MMWR*, 58/16 (1 May 2009), pp. 435–7; "Outbreak of Swine-Origin Influenza A (H1N1) Virus Infection—Mexico, March–April 2009," *MMWR*, LVIII/17 (8 May 2009), pp. 467–70.

2 "WHO Global Influenza Preparedness Plan (2005)," available at www.who.int/en, accessed 7 July 2012.

3 The WHO did itself no favors in this controversy by declining to release

the names of the sixteen influenza experts who had constituted the Emergency Committee for Pandemic H1N1. This group had advised Director-General Chan on the steps the WHO should be taking to respond to the pandemic. See Mark Honigsbaum, "Was Swine Flu Ever a Real Threat?," *Daily Telegraph* (UK, 2 February 2010), available at www.telegraph.co.uk, accessed 18 June 2012; Imogen Foulkes, "WHO Faces Questions over Swine Flu Policy," BBC News Europe (20 May 2010), available at www.bbc.co.uk/news, accessed 18 June 2012; Deborah Cohen and Philip Carter, "WHO and the Pandemic Flu Conspiracies," BMJ, CCCXL (3 June 2010), p. 2912; "WHO Director General's Letter to BMJ Editors" (8 June 2010), available at Media Centre, www.who.int/en, accessed 14 June 2012; and "The International Responses to the Influenza Pandemic: WHO Responds to the Critics" (10 June 2010), available at Global Alert and Response (GAR), www.who.int/en, accessed 14 June 2012. The committee to investigate the WHO response was headed by Harvey Fineberg, who was at that time the president of the Institute of Medicine (USA).

4 For the study see Christine H. Lees et al., "Pandemic (H1N1) 2009: Associated Deaths Detected by Unexplained Death and Medical Examiner Surveillance," *Emerging Infectious Diseases*, XVII/8 (August 2011), pp. 1479–83.

5 See Cecile Viboud, Mark Miller, Donald R. Olson, Michael Osterholm, and Lone Simonsen, "Preliminary Estimates of Mortality and Years of Life Lost Associated with the 2009 A/H1N1 Pandemic in the U.S. and Comparison with Past Influenza Seasons," *PLoS Currents*, XX/2 (March 2010), available at www.ncbi.nlm.nih.gov, accessed 4 June 2012.

6 F. Dawood et al., "Emergence of a Novel Swine-Origin Influenza A (H1N1) Virus in Humans," *New England Journal of Medicine*, CCCLX/25 (18 June 2009), pp. 2605–15; and Vivek Shinde et al., "Triple-Reassortant Swine Influenza A (H1) in Humans in the United States, 2005–2009," *New England Journal of Medicine*, CCCLX/25 (18 June 2009), pp. 2616–25.

7 Gavin J. D. Smith et al., Letters, "Origin and Evolutionary Genomics of the 2009 Swine-Origin H1N1 Influenza A Epidemic," *Nature*, CDLIX (25 June 2009), pp. 1122–5.

8 A promising step in this direction of integrated infectious disease study is the One Health Initiative which seeks to enhance cooperation

and collaboration between physicians, veterinarians, and researchers. Since humans, animals, and the ecosystem are all interlinked, a holistic view of disease is required. See www.onehealthinitiative.com, accessed 18 June 2012.

9 See for example Michael Greger, *Bird Flu: A Virus of our Own Hatching* (New York, 2006), and Mike Davis, *The Monster at our Door: The Global Threat of Avian Flu* (New York, 2005).

Bibliography

"40m Doses of H1N1 Vaccine to be Destroyed," UPI.com, available at www.upi.com, accessed 13 June 2011

Advertisements for "West's Electric Cure," *The Daily Inter Ocean* [Chicago] (1 January 1890), p. 8, collected from *America's Historical Newspaper Database, Newsbank*, accessed 25 August 2008

Albarella, Umberto, Keith Dobney, and Peter Rowley-Conway, "The Domestication of the Pig (*Sus scrofa*): New Challenges and Approaches," in *Documenting Domestication: New Genetic and Archaeological Paradigms*, ed. Melinda A. Zeder, Daniel G. Bradley, Eve Emshwiller, and Bruce D. Smith (Berkeley, CA, 2006), pp. 209–27

Andrewes, C. H., "Epidemiology of Influenza," *Bulletin of the World Health Organization*, VIII (1953), pp. 595–612

Barker, Graeme, *The Agricultural Revolution in Prehistory: Why did Foragers Become Farmers?* (Oxford, 2006)

Barry, John M., *The Great Influenza: The Epic Story of the Deadliest Plague in History* (New York, 2004)

Barua, Dhiman and William Burrows, eds, *Cholera* (Philadelphia, 1974)

Benison, Saul, *Tom Rivers: Reflections on a Life in Medicine and Science* (Cambridge, MA, 1967)

Bentley, Jerry H., *Old World Encounters: Cross-cultural Contacts and Exchanges in Pre-Modern Times* (New York, 1993)

Beveridge, W.I.B., *Influenza: The Last Great Plague; An Unfinished Story of Discovery* (New York, 1977)

——, "The Chronicle of Influenza Epidemics," *History and Philosophy of the Life Sciences*, XIII (1991), pp. 223–35

Bray, R. S., *Armies of Pestilence: The Impact of Disease on History* (New York, 1996)

Breiman, Robert F., Meirion R. Evans, Wolfgang Preiser, James Maguire, Alan Schnur, Ailan Li, Henk Bekedam, and John S. MacKenzie, "Role of China in the Quest to Define and Control Severe Acute Respiratory Syndrome," *Emerging Infectious Diseases*, IX/9 (September 2003), pp. 1037–41

Bresalier, Michael, "Neutralizing Flu: 'Immunological Devices' and the Making of a Virus Disease," in *Crafting Immunity: Working Histories of Clinical Immunology*, ed. Kenton Kroker, Pauline M. H. Mazumdar, and Jennifer Keelan (Aldershot, 2008), pp. 107–44

——, "Transforming Flu: Medical Science and the Making of Virus Disease in London, 1890–1939," PhD Dissertation, Trinity College, Cambridge (2010)

Byerly, Carol R., *Fever of War: The Influenza Epidemic in the U.S. Army during World War I* (New York, 2005)

Cauthen, Angela N., David E. Swayne, Stacey Schultz-Cherry, Michael L. Perdue, and David L. Suarez, "Continued Circulation in China of High Pathogenic Avian Influenza Viruses Encoding the Hemagglutinin Gene Associated with the 1997 H5N1 outbreak in Poultry and Humans," *Journal of Virology*, LXXIV/14 (July 2000), pp. 6592–9

Chan, Paul K. S., "Outbreak of Avian Influenza A (H5N1) Virus Infection in Hong Kong in 1997," *Clinical Infectious Diseases*, XXXIV, Suppl. 2 (2002), pp. S58–S64

Cipolla, Carlo M., *Miasmas and Disease: Public Health and the Environment in the Pre-Industrial Age*, trans. Elizabeth Potter (New Haven and London, 1992)

Cockburn, W. Charles, P. J. Delon, and W. Ferreira, "Origin and Progress of the 1968–69 Hong Kong Influenza Epidemic," *Bulletin of the World Health Organization*, XLI (1969), pp. 345–8

Cohen, Deborah, and Philip Carter, "WHO and the Pandemic Flu Conspiracies," BMJ, CCCXL (3 June 2010)

Cohen, Mark Nathan, *Health and the Rise of Civilization* (New Haven, CT, 1989)

Crawford, Dorothy, *The Invisible Enemy: A Natural History of Viruses* (Oxford, 2000)

Creighton, Charles, *A History of Epidemics in Britain*, vol. I: *From AD 664 to the Great Plague* [1894] (New York, 1965)

——, *A History of Epidemics in Britain*, vol. II: *From the Extinction of the Plague to* the Present Time [1891] (New York, 1965)

Crookshank, F. G., ed., *Influenza: Essays by Several Authors* (London, 1922)

Crosby, Alfred, *Ecological Imperialism: The Biological Expansion of Europe, 900–1900* (Cambridge, 1986)

——, *Epidemic and Peace, 1918* (Westport, CT, 1976), reissued as *America's Forgotten Pandemic* (Cambridge, 1989)

Culbertson, James T., "Plans for United States Cooperation with the World Health Organization in the International Influenza Study Program," *American Journal of Public Health*, XXXIX (January 1949), pp. 37–43

Davenport, F. M., A. V. Hennessy, J. Drescher, J. Mulder, and T. Francis Jr, "Further Observations on the Relevance of Serologic Recapitulations of Human Infection with Influenza Viruses," *Journal of Experimental Medicine*, CXX (1964), pp. 1087–97

——, E. Minuse, A. V. Hennessy, and T. Francis Jr, "Interpretations of Influenza Antibody Patterns of Man," *Bulletin of the World Health Organization*, XLI (1969), pp. 453–60

Davies, Pete, *The Devil's Flu: The World's Deadliest Influenza Epidemic and the Scientific Hunt for the Virus that Caused It* (New York, 2000)

Davis, Dorland J., "World Health Organization Influenza Study Program in the United States," *Public Health Reports*, LXVII/12 (December 1952), pp. 1185–90

Davis, Kingsley, *The Population of India and Pakistan* (Princeton, NJ, 1951)

Davis, Mike, *The Monster at our Door: The Global Threat of Avian Flu* (New York, 2005)

Dawood, Fatima S., Seema Jain, Lyn Finelli, Michael W. Shaw, Stephen Lindstrom, Rebecca J. Garten, Larisa V. Gubareva, Xiyan Xu, Carolyn B. Bridges, and Timothy M. Uyeki, "Emergence of a Novel Swine-Origin Influenza A (H1N1) Virus in Humans," *New England Journal of Medicine*, CCCLX/25 (18 June 2009), pp. 2605–15

De Jong, J. C., E.C.J. Claas, A.D.M.E. Osterhaus, R. G. Webster and W. I. Lim, "A Pandemic Warning?," *Nature*, CCCLXXXIX (9 October 1997), p. 554

Diamond, Jared, *Guns, Germs, and Steel: The Fates of Human Societies* (New York, 1997)

Donaldson, Robert, "The Bacteriology of Influenza: With Special Reference to Pfeiffer's Bacillus," in *Influenza: Essays by Several Authors*, ed. F. G. Crookshank (London, 1922), pp. 139–236

Donkin, R. A., *The Muscovy Duck, Carina Moschata Domestication: Origins, Dispersal and Associated Aspects of the Geography of Domestication* (Rotterdam, 1989)

Dowdle, Walter R., Marion T. Coleman, Elmer C. Hall, and Violeta Knez, "Properties of the Hong Kong Virus: 2. Antigenic Relationship of the Hong Kong Virus Hemagglutinin to that of Other Human Influenza A Viruses," *Bulletin of the World Health Organization*, XLI (1969), pp. 419–24

Duffy, John, *The Sanitarians: A History of American Public Health* (Urbana and Chicago, 1990)

Editorial, *American Journal of Public Health*, VIII (October 1918), pp. 787–8

Etheridge, Elizabeth W., *Sentinel for Health: A History of the Centers for Disease Control* (Berkeley, CA, 1992)

Eyler, John M., "De Kruif's Boast: Vaccine Trials and the Construction of a Virus," *Bulletin of the History of Medicine*, LXXX/3 (Fall 2006), pp. 409–38

——, *Victorian Social Medicine: The Ideas and Methods of William Farr* (Baltimore and London, 1979)

Fedson, David S., Andre Waida, J. Patrick Nicol, and Leslie L. Roos, "'The Old Man's Friend,'" *The Lancet*, CCCXLII/8870 (28 August 1993), p. 561

Fenner, F., D. A. Henderson, I. Arita, Z. Jezek, and I. D. Ladnyi, *Smallpox and its Eradication* (Geneva, 1988)

Fildes, Paul, "Richard Friedrich Johannes Pfeiffer, 1858–1945," *Biographical Memoirs of the Royal Society*, vol. II (November 1956), pp. 237–47

Foulkes, Imogen, "WHO Faces Questions over Swine Flu Policy," BBC News Europe (20 May 2010), available at www.bbc.co.uk/news, accessed 18 June 2012

Francis, Thomas, "Transmission of Influenza by a Filterable Virus," *Science*, n.s. LXXX/2081 (16 November 1934), p. 458

Francis Jr, Thomas, "Vaccination Against Influenza," *Bulletin of the World Health Organization*, VIII (1953), pp. 725–41

——, and Jonas E. Salk, "A Simplified Procedure for the Concentration and Purification of Influenza Virus," *Science*, n.s. XCVI/2500 (27 November 1942), pp. 499–500

——, Jonas E. Salk, and J. J. Quilligan Jr, "Experience with Vaccination Against Influenza in the Spring of 1947: A Preliminary Report," *American Journal of Public Health*, XXXVII (August 1947), pp. 1013–16

Fukumi, Hideo, "Interpretation and Significance of Hong Kong Antibody in Old People Prior to the Hong Kong Influenza Epidemic," *Bulletin of the World Health Organization*, XLI (1969), pp. 469–73

Furminger, Ian G. S., "Vaccine Production," in *Textbook of Influenza*, ed. Karl G. Nicholson, Roger G. Webster, and Alan J. Hay (Oxford, 1998), pp. 324–32

Gadsby, Patricia, "Fear of Flu: Pandemic Influenza Outbreaks," *Discover*, XX/1 (January 1999), pp. 82–9

Gallut, J., "The Cholera Vibrios," in *Cholera*, ed. Dhiman Barua and William Burrows (Philadelphia, 1974), pp. 17–40

Garten, Rebecca J., et al., "Antigenic and Genetic Characteristics of Swine-Origin 2009 A (H1N1) Influenza Viruses Circulating in Humans," *Science*, CCCXXV (10 July 2009), pp. 197–201

Gernhart, Gary, Office of PHS Historian, "A Forgotten Enemy: PHS's Fight Against the 1918 Influenza Pandemic," *Public Health Reports*, CXIV (December 1999), pp. 559–61

Giles-Vernick, Tamara, and Susan Craddock, eds, with Jennifer Gunn, *Influenza and Public Health: Learning from Past Pandemics* (London, 2010)

Greger, Michael, *Bird Flu: A Virus of Our Own Hatching* (New York, 2006)

Grob, Gerald N., *The Deadly Truth: A History of Disease in America* (Cambridge, MA, 2002)

Haaheim, L. R., "Basic Influenza Virology and Immunology," in *Introduction to Pandemic Influenza*, ed. Jonathan Van-Tam and Chloe Sellwood (Oxford, 2010), pp. 14–27

Harper, James, "The Tardy Domestication of the Duck," *Agricultural History*, XLVI/3 (July 1972), pp. 385–9

Hays, J. N., *The Burdens of Disease: Epidemics and Human Response in Western History* (New Brunswick, NJ, 2000)

Henderson, D. A., *Smallpox: The Death of a Disease* (Amherst, NY, 2009)

——, Brooke Courtney, Thomas V. Inglesby, Eric Toner, and Jennifer B. Nuzzo, "Public Health and Medical Response to the 1957–58 Influenza Pandemic," *Biosecurity and Bioterrorism: Biodefense Strategy, Practice, and Science*, VII/3 (2009), pp. 265–73

Heymann, David L., and Guenael Rodier, "Global Surveillance, National Surveillance, and SARS," *Emerging Infectious Diseases*, X/2 (February 2004), pp. 173–5

Hilleman, Maurice, "Six Decades of Vaccine Development: A Personal History," *Nature Medicine Vaccine Supplement*, IV/5 (May 1998), pp. 507–14

Hippocrates, "The Cough of Perinthus," in *Hippocrates*, vol. VII, Book VI: *Of the Epidemics*, trans. and ed. Wesley D. Smith (Cambridge, MA, 1994)

Hirsch, August, *Geographical and Historical Pathology*, vol. 1: *Acute Infective Diseases*, trans. from the 2nd German edn Charles Creighton (London, 1883)

Hirschhorn, Norbert, Nathaniel F. Pierce, Kazumine Kobari, and Charles C. J. Carpenter Jr, "The Treatment of Cholera," in *Cholera*, ed. Dhiman Barua and William Burrows (Philadelphia, 1974), pp. 235–52

Holland, John, Katherine Spindler, Frank Horodyski, Elizabeth Grabau, Stuart Nichol, and Scott VandePol, "Rapid Evolution of RNA Genomes," *Science*, n.s. CCXV/4540 (26 March 1982), pp. 1577–85

Honigsbaum, Mark, "The Great Dread: Cultural and Psychological Impacts and Responses to the 'Russian' Influenza in the United Kingdom, 1889–1893," *Social History of Medicine*, XXIII/2 (August 2010), pp. 299–319

——, "The 'Russian' Influenza in the UK: Lessons Learned, Opportunities Missed," *Vaccine*, 29S (2011), pp. B11–B15

——, "Was Swine Flu Ever a Real Threat?" *Daily Telegraph* (UK, 2 February 2010), available at www.telegraph.co.uk, accessed 18 June 2012

"Hong Kong Battling Influenza Epidemic," *New York Times*, 17 April 1957, p. 3

Hyams, Edward, *Animals in the Service of Man* (Philadelphia, 1972)

Institute of Medicine, *Emerging Infections: Microbial Threats to Health in the United States* (Washington, DC, 1992)

Johnson, N.P.A.S., "The Overshadowed Killer: Influenza in Britain in 1918–19," in *The Spanish Influenza Pandemic of 1918–19: New Perspectives*, ed. Howard Phillips and David Killingray (London, 2003), pp. 132–55

——, and Juergen Mueller, "Updating the Accounts: Global Mortality of the 1918–1920 'Spanish' Influenza Pandemic," *Bulletin of the History of Medicine*, LXXVI/1 (2002), pp. 105–15

Johnson, Steven, *The Ghost Map: The Story of London's Most Terrifying Epidemic and how it Changed Science, Cities, and the Modern World* (New York, 2006)

Jordan, Edwin O., *Epidemic Influenza: A Survey* (Chicago, 1927)

Kendal, Alan P., Gary R. Noble, John J. Skehel, and Walter Dowdle, "Antigenic Similarity of Influenza A (H1N1) Viruses from Epidemics in 1977–1978 to 'Scandinavian' Strains Isolated in Epidemics of 1950–1951," *Virology*, 89 (1978), pp. 632–6

Kilbourne, Edwin, "Future Influenza Vaccines and the Use of Genetic Recombinants," *Bulletin of the World Health Organization*, XLI (1969), pp. 643–5

——, *Influenza* (New York and London, 1987)

——, Catherine Smith, Ian Brett, Barbara A. Pokorny, Bert Johansson, and Nancy Cox, "The Total Influenza Vaccine Failure of 1947 Revisited: Major Intrasubtypic Antigenic Change Can Explain Failure of Vaccine in a Post-World War II Epidemic," *Proceedings of the National Academy of Sciences*, XCIX/16 (6 August 2002), pp. 10748–2

Kolata, Gina, *Flu: The Story of the Great Influenza Pandemic of 1918 and the Search for the Virus that Caused It* (New York, 1999)

Kraut, Alan, *Silent Travelers: Germs, Genes, and the "Immigrant Menace"* (New York, 1994)

Ku, A.S.W., and L.T.W. Chan, "The First Case of H5N1 Avian Influenza Infection in a Human with Complications of Adult Respiratory Distress Syndrome and Reye's Syndrome," *Journal of Paediatric Child Health*, XXXV (1999), pp. 207–9

Langdon, Robert, "When the Blue-Egg Chickens Come Home to Roost: New Thoughts on the Prehistory of the Domestic Fowl in Asia, America, and the Pacific Islands," *Journal of Pacific History*, XXIV/2 (October 1989), pp. 164–92

Langmuir, Alexander, "William Farr: Founder of Modern Concepts of Surveillance," *International Journal of Epidemiology*, V/1 (1976), pp. 13–18

Laver, William Graeme, Norbert Bischofberger, and Robert G. Webster, "The Origin and Control of Pandemic Influenza," *Perspectives in Biology and Medicine*, XLIII/2 (Winter 2000), pp. 173–92

Layne, Scott P., "Human Influenza Surveillance: The Demand to Expand," *Emerging Infectious Diseases*, XII/4 (April 2006), pp. 562–8

Lees, Christine H., et al., "Pandemic (H1N1) 2009: Associated Deaths Detected by Unexplained Death and Medical Examiner Surveillance," *Emerging Infectious Diseases*, XVIII/8 (August 2011), pp. 1479–83

Leffell, Mary S., Albert D. Donnenberg, and Noel R. Rose, eds, *Handbook of Human Immunology* (Boca Raton, FL, 1997)

Lilienfeld, D. E., "Celebration: William Farr (1807–1883), An Appreciation on the 200th Anniversary of His Birth," *International Journal of Epidemiology*, XXXVI (2007), pp. 985–7

Livy, "The Siege of Achradina," in *Livy*, vol. III: *The History of Rome*, 5 vols, trans. George Baker (New York, 1836)

McCracken, Kevin, and Peter Curson, "Flu Downunder: A Demographic and Geographic Analysis of the 1919 Epidemic in Sydney, Australia," in *The Spanish Influenza Pandemic of 1918–19: New Perspectives*, ed. Howard Phillips and David Killingray (London, 2003), pp. 110–31

McNeill, William H., *Plagues and Peoples* (New York, 1976)

——, *The Rise of the West: A History of the Human Community* (Chicago and London, 1963)

Markel, Howard, *When Germs Travel: Six Major Epidemics that have Invaded America since 1900 and the Fears They have Unleashed* (New York, 2004)

Marsh, James A., and Marion D. Kendall, eds, *The Physiology of Immunity* (Boca Raton, FL, 1996)

Martindale, Diane, "No Mercy," *New Scientist* (14 October 2000), p. 2929

Masurel, N., "Serological Characteristics of a 'New' Serotype of Influenza A Virus: The Hong Kong Strain," *Bulletin of the World Health Organization*, XLI (1969), pp. 461–8

Matrosovich, Mikhail, et al., "Early Alteration of the Receptor-Binding Properties of H1, H2, and H3 Avian Influenza Virus Hemagglutinins After Their Introduction into Mammals," *Journal of Virology*, LXXIV/18 (September 2000), pp. 8502–12

Matsui, Akira, Naotaka Ishiguro, Hitomai Hongo, and Masao Minagawa, "Wild Pig? Or Domesticated Boar? An Archaeological View of the Domestication of Sus scrofa in Japan," in *First Steps of Animal Domestication: New Archeozoological Approaches*, ed. J.-D. Vigne, J. Peters, and D. Helmer, Proceedings of the 9th ICAZ Conference, Durham, 2002 (Oxford, 2005), pp. 148–59

Members of the Commission on Influenza, Board for the Investigation and Control of Influenza and other Epidemic Diseases in the Army, Preventive Medical Service, Office of the Surgeon General, United States Army, "A Clinical Evaluation of Vaccination Against Influenza: Preliminary Report," *Journal of the American Medical Association*, CXXIV/14 (1 April 1944), pp. 982–5

Mizelle, Brett, *Pig* (London, 2011)

Molinari, Noelle-Angelique M., et al., "The Annual Impact of Seasonal Influenza in the U.S.: Measuring Disease Burden and Costs," *Vaccine*, XXV (2007), pp. 5086–96

Montefusco, N., in *Riforma Medica*, Naples, XXXIV/28 (13 July 1918), p. 549,

as quoted in "Current Medical Literature," JAMA, LXXI/11
(14 September 1918), pp. 934

Morse, Stephen S., ed., *Emerging Viruses* (New York, 1993)

Mounts, Anthony W., et al., "Case-Control Study of Risk Factors for Avian
Influenza A (H5N1) Disease, Hong Kong, 1997," *The Journal of
Infectious Diseases*, CLXXX (1999), pp. 505–8

Mulder, J., "Asiatic Influenza in the Netherlands," *Lancet* (17 August 1957),
pp. 334

Nelson, Sarah M., ed., *Ancestors for the Pigs: Pigs in Prehistory* (Philadelphia,
1998)

——, "Pigs in the Hongshan Culture," in *Ancestors for the Pigs: Pigs in
Prehistory*, ed. Sarah M. Nelson (Philadelphia, 1998), pp. 99–107

Neustadt, Richard E., and Harvey V. Fineberg, *The Swine Flu Affair: Decision-
making on a Slippery Disease* (Washington, DC, 1978)

Nicholson, Karl G., "Human Influenza," in *Textbook of Influenza*, ed. Karl
G. Nicholson, Robert G. Webster, and Alan J. Hay (Oxford, 1998),
pp. 219–64

——, Roger G. Webster, and Alan J. Hay, eds, *Textbook of Influenza* (Oxford,
1998)

Nicholson-Preuss, Mari Loreena, "Managing Morbidity and Mortality:
Pandemic Influenza in France, 1889–90," MA Thesis, Texas Tech
University (2001)

Offit, Paul, *The Cutter Incident: How America's First Polio Vaccine Led to the
Growing Vaccine Crisis* (New Haven and London, 2005)

——, *Vaccinated: One Man's Quest to Defeat the World's Deadliest Diseases* (New
York, 2007)

Oxford, J. S., "The So-Called Great Spanish Influenza Pandemic of 1918
May Have Originated in France in 1916," *Philosophical Transactions of the
Royal Society of London, Series B*, CCCLVI (2001), pp. 1857–9

——, R. Lambkin, A. Sefton, R. Daniels, A. Elliot, R. Brown, and D. Gill,
"A Hypothesis: the Conjunction of Soldiers, Gas, Pigs, Ducks, Geese,
and Horses in Northern France During the Great War Provided the
Conditions for the Emergence of the 'Spanish' Influenza Pandemic of
1918–19," *Vaccine*, XXIII/7 (January 2005), pp. 940–45

——, A. Sefton, R. Jackson, N.P.A.S. Johnson, and R. S. Daniels. "Who's
that Lady?," *Nature Medicine*, V/12 (December 1999), pp. 1351–2

Parsons, H. Franklin, *Further Report and Papers on Pandemic Influenza*,

1889–92: *Presented to Both Houses of Parliament by Command of Her Majesty* (London, 1893)

——, *Report on the Influenza Epidemic of 1889–90: Presented to Both Houses of Parliament by Command of Her Majesty* (London, 1891)

Patterson, K. David, *Pandemic Influenza, 1700–1900: A Study in Historical Epidemiology* (Totowa, NJ, 1986)

——, and Gerald F. Pyle, "The Geography and Mortality of the 1918 Influenza Pandemic," *Bulletin of the History of Medicine*, LXV (1991), pp. 4–21

Paul, John R., *A History of Poliomyelitis* (New Haven and London, 1971)

Payne, A.M.-M., "The Influenza Programme of WHO," *Bulletin of the World Health Organization*, VIII (1953), pp. 755–74

Peiris, J. S. Malik, Menno D. de Jong, and Yi Guan, "Avian Influenza Virus (H5N1): A Threat to Human Health," *Clinical Microbiology Reviews*, XX/2 (2007), pp. 243–67

Pendergrast, Mark, *Inside the Outbreaks: The Elite Medical Detectives of the Epidemic Intelligence Service* (Boston, 2010)

Perdue, Michael, and David E. Swayne, "Invited Minireview: Public Health Risk from Avian Influenza Viruses," *Avian Diseases*, XLIX/3 (September 2009), pp. 317–27

Phillips, Howard, and David Killingray, eds, *The Spanish Influenza Pandemic of 1918–19: New Perspectives* (London, 2003)

Pollitzer, R., *Cholera* (Geneva, 1959)

Riley, James C., *The Eighteenth-century Campaign to Avoid Disease* (New York, 1987)

Rodriguez-Ocana, Esteban, "Barcelona's Influenza: A Comparison of the 1889–90 and 1918 Autumn Outbreaks," in *Influenza and Public Health: Learning from Past Pandemics*, ed. Tamara Giles-Vernick and Susan Craddock, with Jennifer Gunn (London, 2010), pp. 41–68

Rosen, George, *A History of Public Health* (New York, 1958)

Rosenberg, Charles, *The Cholera Years: The United States in 1832, 1849, and 1866* [1962] (Chicago, 1987)

Salk, Jonas E., "Reactions to Concentrated Influenza Virus Vaccines," *Journal of Immunology*, 58 (1948), pp. 369–95

Scholtissek, C., "Pigs as the 'Mixing Vessel' for the Creation of New Pandemic Influenza A Viruses," *Medical Principles and Practice*, II/2 (1990/91), pp. 65–71

Shanks, Niall, and Rebecca A. Pyles, "Evolution and Medicine: The Long Reach of 'Dr Darwin'," *Philosophy, Ethics, and Humanities in Medicine*, II/4 (3 April 2007), available at www.peh-med.com

Shinde, Vivek, Carolyn B. Fridges, Timothy M. Uyeki, Bo Shu, Amanda Balish, Xiyan Xu, Stephen Lindstrom, Larisa V. Gubareva, Varough Deyde, Rebecca J. Garten, Meghan Harris, Susan Gerber, Susan Vagasky, Forrest Smith, Neal Pascoe, Karen Martin, Deborah Dufficy, Kathy Ritger, Craig Conover, Patricia Quinlisk, Alexander Klimov, Joseph S. Bresee, and Lyn Finelli, "Triple-Reassortant Swine Influenza A (H1) in Humans in the United States, 2005–2009," *New England Journal of Medicine*, CCCLX/25 (18 June 2009), pp. 2616–25

Shope, Richard, "Swine Influenza I: Experimental Transmission and Pathology," *Journal of Experimental Medicine*, LIV (1931), pp. 349–60

——, "Swine Influenza II: A Hemophilic Bacillus from the Respiratory Tract of Infected Swine," *Journal of Experimental Medicine*, LIV (1931), pp. 361–72

——, "Swine Influenza III: Filtration Experiments and Etiology," *Journal of Experimental Medicine*, LIV (1931), pp. 373–85

Shortridge, Kennedy, Peng Gao, Yi Guan, Toshihiro Ito, Yoshihiro Kawaoka, Deborah Markwell, Ayato Takada, and Robert G. Webster, "Interspecies Transmission of Influenza Viruses: H5N1 Virus and a Hong Kong SAR Perspective," *Veterinary Microbiology*, LXXIV (2000), pp. 141–7

Shortridge, K. F., J.S.M. Peiris, and Y. Guan, "The Next Influenza Pandemic: Lessons from Hong Kong," *Journal of Applied Microbiology*, XCIV (2003), pp. 70S–79S

Silverstein, Arthur M., *Pure Politics and Impure Science: The Swine Flu Affair* (Baltimore and London, 1981)

Sims, L. D., "Lessons Learned from Asian H5N1 Outbreak Control," *Avian Diseases*, LI/1, Supplement: Sixth International Symposium on Avian Influenza (March 2007), pp. 174–81

——, and Ian H. Brown, "Multicontinental Epidemic of H5N1 HPAI Virus (1996–2007)," in *Avian Influenza*, ed. David Swayne (Ames, IA, 2008), pp. 251–86

——, and Andrew J. Turner, "Avian Influenza in Australia," in *Avian Influenza*, ed. David Swayne (Ames, IA, 2008), pp. 239–50

——, T. M. Ellis, K. K. Liu, K. Dyrtirg, H. Wong, M. Peiris, Y. Guan, and K. F.

Shortridge, "Session Keynote Address: Avian Influenza in Hong Kong 1997–2002," *Avian Diseases*, XLVII, Special Issue, Proceedings of the Fifth International Symposium on Avian Influenza (2003), pp. 832–8

Smith, Derek J., "Predictability and Preparedness in Influenza Control," *Science*, CCCXII (21 April 2006), pp. 392–4

Smith, Gavin J. D., Dhanasekaran Vijaykrishna, Justin Bahl, Samantha J. Lycett, Michael Worobey, Oliver G. Pybus, Siu Kit Ma, Chung Lam Cheung, Jayna Raghwani, Samir Bhatt, J. S. Malik Peiris, Yi Guan, and Andrew Rambaut, Letters, "Origin and Evolutionary Genomics of the 2009 Swine-Origin H1N1 Influenza A Epidemic," *Nature*, CDLIX (25 June 2009), pp. 1122–5

Smith, Richard D., "Responding to Global Infectious Disease Outbreaks: Lessons from SARS on the Role of Risk Perception, Communication, and Management," *Social Science and Medicine*, LXIII/12 (2006), pp. 3113–23

Stobbe, Mike, "Millions of Vaccine Doses to be Burned," 1 July 2010, Associated Press Online, accessed via LexisNexis Academic, 30 August 2010

Stratton, Kathleen, Donna A. Alamario, Theresa Wizemann, and Marie C. McCormick, eds, *Immunization Safety Review: Influenza Vaccines and Neurological Complications* (Washington, DC, 2004)

Subbarao, Kanta, Alexander Klimov, Jacqueline Katz, Helen Regnery, Wilina Lim, Henrietta Hall, Michael Perdue, David Swayne, Catherine Bender, Jing Huang, Mark Hemphill, Thomas Rowe, Michael Shaw, Xiyan Xu, Keiji Fukuda, and Nancy Cox, "Characterization of an Avian Influenza A (H5N1) Virus Isolated from a Child with a Fatal Respiratory Illness," *Science*, CCLXXIX/5349 (16 January 1998), pp. 393–6

Taubenberger, Jeffrey K., and David M. Morens, "1918 Influenza: The Mother of All Pandemics," *Emerging Infectious Diseases*, XII/1 (January 2006), pp. 15–22

——, Ann H. Reid, Thomas A. Janczewski, and Thomas G. Fanning, "Integrating Historical, Clinical, and Molecular Genetic Data in Order to Explain the Origin and Virulence of the 1918 Spanish Influenza Virus," *Philosophical Transactions: Biological Sciences*, CCCLVI/1416 (29 December 2001), pp. 1829–39

——, Ann H. Reid, Amy E. Krafft, Karen E. Bijwaard, and Thomas

G. Fanning, "Initial Genetic Characterization of the 1918 'Spanish' Influenza Virus," *Science*, CCLXXV (21 March 1997), pp. 1793–6

Thompson, Theophilus, *Annals of Influenza or Epidemic Catarrhal Fever in Great Britain from 1510 to 1837* (London, 1852)

Thorne, R. Thorne, "Introduction by the Medical Officer, to Local Government Board," in H. Franklin Parsons, *Further Report and Papers on Pandemic Influenza, 1889–92: Presented to Both Houses of Parliament by Command of Her Majesty* (London, 1893), pp. i–xi

Tomes, Nancy, *The Gospel of Germs: Men, Women, and the Microbe in American Life* (Cambridge, MA, 1998)

Torrey, E. Fuller, and Robert H. Yolkey, *Beasts of the Earth: Animals, Humans and Disease* (New Brunswick, NJ, 2005)

Townsend, John F., "History of Influenza Epidemics," *Annals of Medical History*, V/6 (November 1933), pp. 533–47

Tyrrell, David, "Discovery of Influenza Viruses," in *Textbook of Influenza*, ed. Karl G. Nicholson, Robert G. Webster, and Alan J. Hay (London, 1998), pp. 19–26

Vaughan, Warren T., *Influenza: An Epidemiologic Study* (Baltimore, MD, 1921)

Viboud, Cecile, Rebecca F. Grais, Bernard A. P. Lafont, Mark A. Miller, and Lone Simonsen, "Multinational Impact of the 1968 Hong Kong Influenza Pandemic: Evidence for a Smoldering Pandemic," *Journal of Infectious Diseases*, CXCII (15 July 2005), pp. 233–48

Viboud, Cecile, Mark Miller, Donald R. Olson, Michael Osterholm, and Lone Simonsen, "Preliminary Estimates of Mortality and Years of Life Lost Associated with the 2009 A/H1N1 Pandemic in the U.S. and Comparison with Past Influenza Seasons," *PLoS Currents*, XX/2 (March 2010), available at www.ncbi.nlm.nih.gov, accessed 4 June 2012

Vigne, J.-D., J. Peters, and D. Helmer, eds, *First Steps of Animal Domestication: New Archeozoological Approaches*, Proceedings of the 9th ICAZ Conference, Durham, 2002 (Oxford, 2005)

Webster, Robert G., "William Graeme Laver," *Biographical Memoirs of Fellows of the Royal Society*, available at http://rsbm.royalsocietypublishing.org, accessed 5 June 2011

——, and William J. Bean Jr, "Evolution and Ecology of Influenza Viruses: Interspecies Transmission," in *Textbook of Influenza*, ed. Karl G. Nicholson, Robert G. Webster, and Alan J. Hay (Oxford, 1998), pp. 109–19

Webster, R. G., K. F. Shortridge, and Y. Kawaoka, "Influenza: Interspecies Transmission and Emergence of New Pandemics," FEMS *Immunology and Medical Microbiology*, XVIII (1997), pp. 275–9

Wilschut, Jan C., Janet E. McElhaney, and Abraham M. Palache, *Influenza*, 2nd edn (Edinburgh, 2006)

Winn Jr, Washington C., "Influenza and Parainfluenza Viruses," in *Pathology of Infectious Diseases*, vol. 1, ed. Daniel H. Connor (Stamford, CT, 1997), pp. 221–7

Winslow, C.E.A., *Man and Epidemics* (Princeton, NJ, 1952)

Wise, Darla J., and Gordon R. Carter, *Immunology: A Comprehensive Review* (Ames, IA, 2002)

Wolbach, S. Burt, and Channing Frothingham, "The Influenza Epidemic at Camp Devens in 1918: A Study of the Pathology of the Fatal Cases," *Archives of Internal Medicine*, XXXII/4 (October 1923), pp. 571–600

Wood, John M., and Michael S. Williams, "History of Inactivated Influenza Vaccines," in *Textbook of Influenza*, ed. Karl G. Nicholson, Robert G. Webster, and Alan J. Hay (Oxford, 1998), pp. 317–23

Acknowledgments

No historian can conduct his work without the talents and efforts of a number of other people who aid him in his task. I am therefore delighted to have the opportunity to acknowledge the efforts and expertise of others who have helped me along the way in writing this book. Wichita State University provided a congenial setting for research and writing and I greatly appreciate the funding provided by the University that allowed me to travel to Geneva to visit the World Health Organization Archives and for travel to Beijing to present a paper at the World History Conference. This paper eventually served as the heart of chapter Five. I also want to thank Marie Villemin at the WHO Archives for her efficient and gracious assistance during my research trip to Geneva, and the kind folks at Ablah Library (WSU) both in collections and in Interlibrary Loan for quickly tracking down the more obscure books and articles I requested.

One of the great pleasures in conducting my research on influenza over the years has been the opportunity to speak with some of the leading lights in scientific and public health research on the influenza virus. Almost without exception, I have found these experts to be extraordinarily generous with their time and knowledge. I especially want to thank Kennedy Shortridge and Les Sims for their patience and voluminous responses to my queries on events in Hong Kong in 1997, and Claude Hannoun, with whom I had some between-session chats at the marvelous "After 1918: History and Politics of Influenza in the Twentieth and Twenty-First Centuries" conference in Rennes.

I also want to thank the people at Reaktion Books. The editing staff saved me from some embarrassing errors and for that I am grateful. All errors that remain are, of course, my own damn fault. This project began as a suggestion from the Publisher, Michael Leaman, and I appreciate his support, confidence, and most of all patience as this work came together. Finally I am glad to acknowledge Jodi and my boys Brendan, Patrick, and Sean. Without their love and support I simply would be unable to embark on these journeys.

Index